Harvest Journal

Memoir of a Minnesota Farmer
Part I: 1846–1903

Sandra K. Wilcoxon
Frederick A. Cummings

To Uncle Maynard, with best wishes, —Sandra K. Wilcoxon

HATS
OFF

Harvest Journal: Memoir of a Minnesota Farmer, Part I: 1846-1903

Copyright © 2000 Sandra K. Wilcoxon

An excerpt of this book was previously published as "Mr. Krueger's Woman" in *Daughter of Dangerous Dames* by 11th Hour Productions, May 2000.

International Standard Book Number: 1-58736-023-3
Library of Congress Card Number: 00-111229

Published by Hats Off Books
601 East 1st Street, Tucson, Arizona 85705, U.S.A.
www.ipublisher.com

Cover and book design by Atilla L. Vékony.
Cover photo: *Vinton and Elvyn Cummings during haying season, c. 1920 (?).*

Printed in the United States of America

Table of Contents

Foreword

The memoir is based on my great-great-grandfather's journals, which he kept from 1868 through 1937. Fred Cummings was a simple farmer his entire life, but he had an inquisitive mind and a sharp wit. He worked incredibly hard to support his family and keep the farm going. He subscribed to at least seven newspapers, read every book he could get his hands on, and recorded his thoughts, poems and events of the day.

It is fascinating to see what has and has not changed since that time. Events and technology may be different now, but overall themes affecting politics, attitudes and issues are remarkably comparable to those of today. As we face the challenges of the future, we may learn something from the past.

This is a cross-generational project—most of the words are his, most of the organization and punctuation are mine. I have added details about events that he mentioned in passing, and I have edited out a lot of weather and crop reports, accounting details, recipes, and peripheral characters.

Many friends and family members have been supportive of this work. This book is especially a tribute to my grandmother, Lola, Phoebe Myrtle's daughter. She started me on this journey by sharing two big, hand-written volumes of the journal. I am also grateful to Esther Heinzelman, Monica Ford, Phyllis Utley, David Steere and other family members who were keepers of the legacy and shared materials and memories. Special thanks also go to Howard Sandborn and the other volunteers at the genealogy department of the Fillmore County History Center for their tireless assistance with research.

Sandra K. Wilcoxon

HARVEST JOURNAL

Memoir of a Minnesota Farmer
Part I: 1846–1903

In the year A.D. 1906 my daughter Myrtle, before moving to the west, gave me this book for the purpose of a Life Journal. Its original object, as a church register, was something far different. But, as it failed in its first mission, I deem it no robbery to appropriate it to my humble use.

—Frederick Augustus Cummings
Born July 22nd, 1846
In Orange County, Vermont

A Record of Events

The past is gone and memory keeps
Uncertain watch o'er him who sleeps.
Unlooked-for Death may snatch away
A witness for a future day.

The Present's here, oh let it then
Be noted down by Wisdom's pen,
That when it's past, should memory fail,
We'll have a friend to tell the tale.

Vermont to Minnesota

Imagine a six-year-old boy, the sixth of nine children, learning that his mother had died. These are the circumstances that Fred Cummings found himself in, at a time when fathers were not expected to care for their children on their own. They either remarried, or parceled the children out to relatives who would care for them. Fred's father sent his children off to different relatives, even splitting up a set of twin girls who were less than one year old. Eventually his father remarried, but the children were never reunited as a family. This "motherless child" feeling resurfaces in Fred's journals throughout his life and plays a significant role as he looks back to his boyhood days.

About my first recollection worthy of note was the sickness and death of my mother, which means more to me now than I knew of then. Poverty was my inheritance; honesty my only recommendation. With these, at the age of seven, I was sent to the home of my mother's sister. These memories are fixed on my mind the more rigidly because I was homesick whenever I had to move, and I feared and trembled that every stranger who came by the house was going to carry me off with him. This, however, proved my permanent home and the move from Vermont to Minnesota my last move. In the spring of 1855, with a younger brother and a three-year-old baby sister, I came with Aunt and Uncle Fowler and a few neighbors to "Waukokee," south of Rochester, Minnesota. Tired—with a week's journey on rail cars to Galena, Illinois, then steamboat to McGregor, Iowa, and lumber wagon the rest of the way—I remember keeping awake in fear of getting left and lost, until exhausted nature gave way and I became as helpless as may be. When I was taken off the boat at McGregor they carried me to the hotel and put me to bed. I knew what they said to me but I had no power of locomotion whatever.

The story of the first part of our journey might be lengthened out into incident and accident but, when all told, would only be a repetition of the experience of all travelers. However, our ride up the river from Galena, which should have been the most enjoyable part of the journey, was made miserable by the presence of cholera and death on board which the crew tried to deny, but only succeeded in making the passengers distrust. They stopped by an island to bury the dead, not being permitted to carry the disease ashore, then the secret leaked out and gloom and consternation was the result. For myself, I dared not touch a morsel of food or drink water. I remember still the warning of one hypochondriac who was apparently thoroughly frightened, "The less you eat and drink on this boat the better!" A lady passenger entertained us by singing "Lillie Dale" and some other songs that I have forgotten.

In passing through Burr Oak, Iowa, one of our party bought something in a pint bottle that should have been rum, but upon trial he pronounced it "nothing but rot-gut whiskey." The first time I ever heard the name or saw the article. Rum is the beverage of all well-regulated Vermont tipplers when they have passed the hard cider stage and I had heard the fame thereof. The disgust of the rum-soaked Vermonter at being sold the first thing, impressed me with the thought that whiskey must be a terrible thing and the seller of it a great sinner. Early impressions are lasting and my mind still holds the doctrine. I have kept about as shy of it as I would a mad dog.

We were two days on the road from McGregor. Those days, like the last one on the boat seemed like Sunday. So many Sundays in one week rather mixed dates in regard to our arrival. But my memory of events is that we started from Vermont on April 23rd and ended our journey May 1, 1855.

Waukokee was a wilderness indeed. It contained about eleven families and as many shanties of greater or less pretensions. Uncle Fowler started on a land hunt immediately and settled on a place a mile south claimed by one Mr. Howell, who convinced him that he could transfer his especial right to him, which would enable Uncle Fowler to obtain said land (school land settled by Howell before the government survey) at the same rate of other public land. Uncle Fowler paid him $800 for his claim, but found his mistake when he went to the land office—he had to swear that he was the actual settler. But he stayed right by it and paid $5 per acre for 150 acres. Howell moved out, and we moved in, about a week after we "landed."

No use describing the shanty except to say it was one of the poorest shacks on the creek—snakes, toads and gophers came in and out at will. Rain dripped through the shake roof in a score of places, and winter was a terror never to be forgotten. This we endured two years. When a saw mill was put to work at Waukokee, we built a frame house, sixteen by twenty-six feet, all of native timber. Shingles were shaved out of oak for the roof and made a nice tight cover, which was much appreciated. Then the sense of destitution and homesickness vanished and I, for one, began to love my home and surroundings.

— o —

School Days

Fred was a serious scholar, shy, and slight of build. His aunt and uncle were very strict and rigorous disciplinarians. His spare time was spent in the field or reading. Fred remembers every teacher of his youth and revered books, yet economic pressures and the need to help on the farm prevented him from taking his formal education further. Even so, he manages to teach in a country school following in the footsteps of some of his boyhood heroes.

In the spring of 1856 the settlers rallied on behalf of the rising generation and, forming themselves into a school district, each land owner was assessed two house logs to be hauled to the proposed school site—a bee was called for and a house erected. An old-fashioned sawmill down on the William Miller place furnished lumber for the roof. A school was organized straight away with Jerusha Thatcher as teacher, and there I took up again the broken thread of an education began in Vermont at so early an age that I have forgotten the time. Suffice to say I had a Town's *Fourth Reader* to begin with in Minnesota, Webster's old blue speller, and I don't know what other books, if any. Jerusha taught three successive summer terms at Waukokee and I attended every term. Sander's books were obtained for the second term and I had writing and mental arithmetic added to my studies.

The first winter term was undertaken by Martin Kingsbury, during which time I studied and probably mastered the multiplication table, for he drilled us most thoroughly. Then John Cunningham took the winter terms for three straight "heats." He added geography, a double drill in mental arithmetic, and a sort of spelling book drill that I cannot describe. All things were thoroughly done under his administration; even the floggings were long remembered by those who were parties to them.

In 1858 I read the history of the United States as far down as the 1812 war. 'Twas an old, finely-printed book, brought from Vermont, whose author I did not take pains to remember. But, as for making history or physiology a study, it never entered my mind until after my school days had passed. I also read Josephus about that time, and several novels of Indian wars by J. Fenimore Cooper. In fact, I never lacked for reading matter when work was done and probably read some books and papers that were better left alone.

Jimmie Underwood taught winter term at school during 1861–62. I remember the way he drilled us on the words spelled and defined in Sander's *Fourth Reader*. Winter added the Tinkelpaugh boys and Albert and Helen Powers to our school. In 1862–63 Isaac Bagnell held forth and was a dismal failure. W. V. Tunstall, our next teacher, was a refugee from Texas during the war. His hobby was spelling, and it was he who insisted on my having a grammar book, "Seventeen years old and never studied it! Man alive, what have you been doing? Go straight off and get one!" There was no appeal from Tunstall's ultimatum, and grammar at this late hour was added to my studies. Tunstall returned to his Southern home after the war and we heard from him through a newspaper article in which he described the destitution and misery caused by the war. He finished the letter with Jeremiah's Lamentation:

> Oh that my head were waters, and mine eyes
> Were fountains, pouring like the liquid skies,
> Then would I give the mighty floods release
> And weep a deluge for the human race.

Charley Crosby took up the thread for the winter term of 1864–65. Arithmetic was his hobby, and Robinson's *Practical* was the book we worked and worked hard. Helen Powers and I were the senior class and I don't think one beat the other very much, we kept about even all through. But the next term Helen was gone to Decorah, Iowa. She came back and took the next and last term I ever attended and plainly had the best of me.

Whatever may be said of my early training, my religious education was not neglected. Sunday school was one of the first things thought of in the new west, and Waukokee Sunday school was one of the best as early as 1857 and I was always there with a lesson—verses of the New Testament learned "by heart." I could repeat whole chapters then in all four Gospels. Whether I can now I will not say, or whether I profited by what I learned I will not pretend to affirm or deny. At that early day three denominations held forth, and three preachers lived at Waukokee so that every Sunday was preaching day. But, in spite of all their terrible thunderings of the horrors of hell, I wandered away at the age of twenty years first into Universalism, then spiritualism with its attempts to contact the dead, then into bitter repentance at the age of forty, and back into the Methodist Church. However, the church's doctrine of devils and torture have no place in my personal creed—it doesn't harmonize in my mind with eternal justice, wisdom, and mercy, so I just drop it out and say, "Thy way, oh God, not mine."

— 0 —

Boyhood Work

In the spring of 1858 we raised a barn frame, thirty by forty feet, on the place, all of native lumber and shingles. I helped carry up the shingles when out of school. But the part I took in it that stays on memory's page was my going around "warning" out help for the raising, and then taking two dollars and a two-gallon jug and going up on Windy Ridge to Old Doc Taylor's for that terrible stuff that aforetime had so disgusted our tippling tenderfoot from Vermont: rot-gut whiskey. The old man filled the jug, and then filled my soul with dismay by demanding $2.50 when I only had $2. But he finally fixed it by taking what I had and charging the balance, then he strapped it on my shoulder and assured me that it would be heavy before I got home; and it was, both with avoirdupois and shame.

It was somewhere near this date that I took my first practical lesson in physiology and hygiene with special reference to stimulants and narcotics. A neighbor boy came to our house smoking a pipe, and seemed to enjoy it so much and looked so majestic that I thought I must learn the art. There were three things he could do that I strove with might and main to imitate: swim, smoke, and walk a high beam in the barn.

The ambition to imitate those whom I considered my superiors was my failing and often worked a world of mischief and shame to me. But, on this occasion, it was the indirect means of teaching me the great lesson that all poison drugs work injury to the human system, even when taken in small quantities. I presume I puffed at that boy's pipe for the space of ten minutes gaining confidence at every "draft" when, suddenly, a red-hot bolt seemed to shoot down my throat and all was dark. I just knew enough to hand the pipe out in the direction of its owner and then sank unconscious in the snow. When I next saw the light my brother Moses was trying

13

to get me up, and the owner of the mischief was laughing, "ready to split." I, to be smart, got up and called for the pipe, which he readily handed me but, before it touched my lips, I went down again and when next I recovered consciousness was only too anxious to get home and to bed. Tobacco had the "belt" and I was sick unto death. All that day my every breath came as a blast from a fiery furnace. Hell, perhaps, would be a fair description of my agony, but I had it all to myself. No one had the least sympathy for me. I have made many good resolutions and broken many in my time, but I made one that day that stands the test of time and temptation. Not so much perhaps from inherent virtue as from the fact that my nature revolted. I never after could tolerate the vile drug, and for more than a year the sight or smell of an old tobacco pipe would make me sick and feel faint.

In 1859 it was decided that my time was worth more at home than at school during the spring and summer. A decision with which I cheerfully agreed, and I went into the field at whatever work I was set at — yoking and driving oxen after hunting them out of the wilds, dripping with dew, was one of the tasks I will remember. On one such occasion, in chasing a refractory steer, I drove a stub almost through my bare foot. I stopped long enough to remove the thing and then renewed the chase bringing the said steer up, put the yoke on him and drove over to Camp Creek when I found myself unfit for duty and was sent home three miles afoot.

Breaking land was the order of the summer, but when we had the little field on the hill cleared for action, on the 11th of June, our house burned to the ground. This changed my work but did not make it easier. I don't know how much others lost but my own seemed too much to stand up under and I just broke down, that is the size of it. School books, treasures saved from earliest childhood brought from Vermont, and playthings acquired here by long, tedious toil and planning — all gone in an hour. Boys of today can't realize what it was to own a knife or pair of skates at that time. I don't remember how I got my first skates but I do know they cost me four times as much as they do the boy of today, and didn't come down the chimney, either! I believe I managed to own another pair, but my precious school books that I had saved so carefully never were to be mine again. Others took their place to be sure, but here I must claim to differ from the average school boy. I never wore out a book in my life; leaves, cover and all were good when I laid them by for new ones. Even the slate (*that* escaped the fire) lasted me from first to last—I never had but the one. Well, the neighbors turned out with ax and saw and had another frame up by the Fourth of July, and we were living in a new house in September—far from completion, to be sure, but we stayed.

During the summer we lived in a little log cabin a half-mile south on the creek bottom, and on this cabin hangs a tale. The biggest flood ever known at that time came upon us the third of July and came near floating us all away. It rained some that night, and many lost their lives in apparently safer places than we occupied. Everything below the rafters was afloat except the stove. I don't know what loss we sustained, except Uncle Fowler lost his pocketbook with all his available cash in it, but later found it in the cellar hole when the water was gone. This, with the big scare over the great comet and the "John Brown raid" finished the excitement for that year. John Brown was Aunt Lydia's relation and they were agitated over the treatment he received.

After considerable bother and several unsuccessful trials with Norwegian oxen, Uncle Fowler bought from a traveling herd two cows and a yoke of big oxen. I remember well the trade he made. It was a big herd of cows and oxen—good, bad and indifferent—and to pick among them was no small task. The cows finally selected were appraised "$30 for one and $40 for t'other," and for the oxen he paid $125. We drove them home after a long chase for one of them and then our business was to bring home the cattle from the wilds. Moses and I—seven and nine-year-old boys—made many crooked paths and unsuccessful sallies. But the rod and reproof worked wonders with us and we finally developed into first-class cowboys. The woods were full of cattle and bells, but we could trace out our own among them more accurately than anyone in sight. Our musical talent was thus early brought in play and trained to a purpose; we just had to know our bell and the least tinkle sent us off through brush and over rushing waters at the risk of snakes or anything else we might meet.

Only once do I remember any serious mishap. I came near drowning trying to wade the swollen creek and was saved by the outcry of my brother. It seems to me that I was tossed up and down and tumbled about a dozen times in that swift current in a semiconscious state, rocked to sleep and perfectly happy when under water, then aroused to a sense of my situation by the screams and cries of my brother when I came to the surface. At last, carried by the rushing tide under some overhanging willows and hearing the shout "Catch hold of the bush!" repeated in frantic despair, I made one last effort and held myself there when, as I suppose, the water pushed me onto the shore and Moses pulled me to dry land. For I never really knew how it happened; my first recollections were of belching water and saying all manner of wild and foolish things.

How came I to get into the water? I waded deliberately in, with the understanding that if I made the "riffle," Moses was to follow. It was not the first time I had tried to cross that day but, here, the force of the current proved my undoing. The water in those days was never roily, even at flood tide we could see the bottom. Wet? Well, yes, wading in the creek to our necks would make us no more so, but that never disturbed our minds in the least.

These are but incidents, by the way, and not any reflection on anyone. I am finding no fault that I was trained to work and hunt cattle. The steady work was my salvation and I am not sorry for the severe discipline that I was subjected to.

What I am sorry for is this: being told that I had schooling enough for any ordinary boy, and that Abraham Lincoln never had my chance for schooling. That I never would need grammar, unless I expected to be a lawyer or a preacher. That singing school never would do me any good. In fact, all the encouragement I received tended to make me suppose that I had learned all there was worth my while. This makes me sorry now. I never had courage to ask for those "useless" books, and so went without until I could buy them for myself. They did buy me a big McNally's geography book for which I was very thankful. I have it yet—battered and worse for its forty years of wear but it is all there, I think. Now, I am not complaining of this. I just want to emphasize the idea of showing children their need of study—don't tell your boy how smart he is, how fast he learns, or impress him with the notion that he knows more than some whole families—either to get work out of him or for any other purpose. This is not a rare thing and it is a most stupid blunder, not to say a crime. We see the effects of it everywhere we go in wrecked, disappointed lives who have found too late that they are poorly equipped for the stern realities of life.

— o —

Civil War

Several major themes that recur throughout Fred's life become evident here. Patriotism and respect for military leaders, spiritual exploration and reflection, a love of music and, always, a focus on family and farm matters.

The War of the Rebellion came in time to count me out so far as soldiering is concerned. I was obliged to stay home much against my will. Like many another thoughtless youth I imagined that to enlist and go with the army would make a hero of me. But the recruiting officer looked at my small stature and said, "Men are what we want." That settled me, and I tried to think I could be a hero at home by waiting on the lonely, almost helpless women left to care for themselves while their sons or husbands were gone. I cut all of one woman's wood for a year and waited on several others more or less. In fact, all who stayed at home were kept as busy as beavers winter and summer caring for the "war widows," so that I was only one among the number.

The War of the Rebellion will ever be fresh in my memory. I had four brothers in the army. Anxiety was no name for our suspense and doubt and fears. News of battle came every week, and often every day for a week at a time until the dead were past counting. Both sides claiming the victory, and neither side knew the truth of the matter till after the war was over and impartial history set us right. It was during this war period, in February 1864, that I was stricken down with diphtheria. This terrible disease raged through the land like pharaoh's plague until there was mourning in every house. I barely escaped with my life and didn't do a stroke of work until harvest. That fall was the last call for soldiers and I wanted to go with the Eleventh Minnesota Volunteers but had not sufficiently recovered from my tussle with disease to bear inspection. Thus ends my war record. But when the soldiers returned — those who lived — with their tales of suffering in battle, on the march, the tent life, the rebel prison pen, and the abusive treatment of their own officers, I was content to be only a boy doing chores for the women.

Phrenology, a mingling of truth and error, interested me much along in the early 1860s. Delineating a man's character by bumps and facial expressions was about as satisfactory as the prognostications of the long-distance weather prophet — there were too many complications and accidents of nature to be considered. Professor C. S. Powers lived in our area and lectured about the country. I attended several of his meetings and he lent me books and magazines. He sold me a chart of my peculiarities that I prized highly and studied very thoroughly, but I finally came to the conclusion that there was just truth enough in it to make an interesting study and humbug enough to suit the average mortal.

I also saw a phenomenon called "spirit rappings." I was well-satisfied that certain persons have an unknown power similar to mesmerism, which by laying the hand on a table or other like structure caused it to move without using any physical strength. Yet I cannot lay it to any supernatural power, for I believe that for everything under the sun there is a natural cause.

It was about this time, in 1866, that I took my first lessons in vocal music. A singing school was organized at the schoolhouse and, under the tuition of one Smith, a class of us began the study of do, re, mi. Our teacher was an old man with a very unmusical voice and, I suspect, not much talent, but we made quite satisfactory progress for Professor Powers and William Wilber made up wherein Smith lacked and we had some fine voices in the class. Not to brag of my own, but I enjoyed that winter and that work more than I shall try to express. Music, either instrumental or vocal, ever had an indescribable fascination for me and at times led me into several ridiculous circumstances. As a small boy I used to creep up under Aunt Lydia's window to hear her sing — she had a charming voice — and as a grownup I stood spellbound in front of a neighbor's house to catch the mellow notes of a melodeon. One time, I ran down the streets of Fountain after a band only to return disgusted and ashamed of myself — it was in a saloon. These are only a few of the fool things I have done — some are too ridiculous to mention. I learned to my grief that music, although of heavenly origin, may be prostituted to most unholy purposes.

In April of 1866 my brother Frank came here from Vermont, fresh from the Army of Virginia, where he served under General George A. Custer. He remembered well a speech his commander gave at Appomattox, near the close of the war, and sent me a copy of it:[*]

Headquarters, Third Cavalry Division;
Appomattox C. St. Va., April 9th, 1865

Soldiers of the Third Cavalry Division:

With profound gratitude toward the God of battles, by whose blessings our enemies have been humbled and our arms rendered triumphant, your commanding General avails himself of this, his first opportunity to express to you his admiration of the heroic manner in which you have passed through the series of battles which today resulted in the surrender of the enemy's entire army.

The record established by your indomitable courage is unparalleled in the annals of war. Your prowess has won for you even the respect and admiration of your enemies. During the past six months, although in most instances confronted by superior numbers, you have captured from the enemy, in open battle, one hundred and eleven pieces of field artillery, sixty-five battle flags, and upwards of ten thousand prisoners of war including seven general officers. Within the past ten days and included in the above, you have captured forty-six pieces of field artillery and thirty-seven battle flags. You have never lost a gun, never lost a color, and have never been defeated; and notwithstanding the numerous engagements in which you have borne a prominent part, including those memorable battles of the Shenandoah, you have captured every piece of artillery, which the enemy has dared to open upon you. The near approach of peace renders it

---

[*]   My source for this speech is a sheaf of notepaper in unknown handwriting, folded as if it had been in an envelope. The speech was published in 1876, and Fred's brother Frank or another family member probably copied it out and sent it to him.

improbable that you will again be called upon to undergo the
fatigues of the toilsome march, or the exposure of the battlefield,
but should the assistance of keen blades wielded by your sturdy
arms be required to hasten the coming of that glorious peace for
which we have been so long contending, the General
commanding is proudly confident that in the future as in the
past, every demand will meet with a hearty and willing response.
Let us hope that our work is done, and that, blessed with the
comforts of peace, we may be permitted to enjoy the pleasures of
home and friends. For our comrades who have fallen, let us ever
cherish a grateful remembrance. To the wounded and to those
who languish in Southern prisons, let our heartfelt sympathy be
tendered. And now, speaking for myself alone, when the war is
ended and the task of the historian begins — when those deeds of
daring which have rendered the name and fame of the Third
Cavalry Division imperishable are inscribed upon the bright
pages of our country's history, I only ask that my name be
written as that of the commander of the Third Cavalry Division.

> — G. A. Custer
> Brevet Major General Commanding
> Official: L. W. Barnhart, Captain and A.A.A.G.

General Custer was a true leader of men, greatly admired by
all who knew him or heard of his feats. Proud as we were of
brother Frank's service in his division, we were glad and relieved
that he survived the strife and trials of the war.

That year, brother Moses rebelled against home rule and
started life on his own. I began dabbling in poetry and keeping an
account of passing events. John, my oldest brother, had come here
five years earlier with a wife and two small children. They proved
a never-ending source of annoyance to me. The story of that family
is a sad warning to others who venture too close to the realms of
evil abandonment. The wife finally committed suicide in a fit of
insanity, and the children went to the bad; some suicided, some by
alcohol. I close my eyes on the scene — God pity them.

Owing to the heavy failure of crops in 1866 all produce was
almost used up by May of 1867, and vegetation was starting very
slow due to cold weather. Produce was selling for very high prices,
$1 per bushel for corn and oats, and wheat at $2, but it was very
scarce if it could be found at all. By June, grain on the bottom land
was very promising, with wheat looking thick and thrifty, but

large numbers of bug eggs were on the potato plants by the time they were four inches high, and they suffered greatly.

We had much heavy thunder and many casualties from lightning in July. It was the most vivid lightning I ever saw. One flash darted three ways at once, the next darted across the sky and then broke in the center with half of it curling up like hair when thrown on a red-hot iron. These were followed by crashes of thunder that shook the ground and made the windows jar.

My 21st birthday came on the 22nd of July, 1867. Uncle Fowler gave me a colt and promised to buy a mate for it at the end of the season for my work, which he did. I took another long term at singing school that summer and into the fall taught by Aaron Dayton, the memory of which is a sweet inspiration for I enjoyed it beyond measure. I thought then that Aaron was a perfect teacher as well as a noble-hearted man, and the class — their memory sends the sweet-sad floods of vain longing for bygone days to my heart as nothing else has done.

The old log schoolhouse gave place to a new stone one that year, built at a cost of $1,300; too late for me to ever attend school in it. I was induced to try my hand at teaching the following winter and now I wonder at my audacity. Still, I gave so good account of myself that they hired me for four winter terms straight along, so I was not the only fool on the job.

— o —

Impeachment

Early in 1998, the United States was rocked by political scandal involving the impeachment of the president. Many of us watched these proceedings disgusted by both the moral issues involved and the partisan process. One hundred and thirty years earlier, the nation followed with interest the impeachment proceedings against President Andrew Johnson. Fred was teaching school and working on Uncle Fowler's farm, and in November took great pride in casting his ballot for the first time in his life.

I had a very pleasant time teaching school over the next three winters but I worked under difficulties. It was hard to keep the schoolhouse warm, and many students had to share books, there not being enough to go around. I had an average of sixteen scholars in a twelve by fourteen-foot schoolhouse but, notwithstanding, we had very good order. Some few came to make disturbance, but every Norwegian was orderly and studious. The school paid $25 per month for three months. I lived at Uncle Fowler's, walked a mile and a half to school, and enjoyed my work. No work seemed irksome to me then, and teaching was a pleasure always. When the school term closed in the spring I began work for Uncle Fowler in the fields.

During early 1868, I busied myself helping Uncle Fowler cutting firewood, making rails, plowing, and hauling manure. We got thirty-one acres of small grain in the ground in April, the grass looked green, and no one complained of lack of fodder that spring. I owned one sheep and two lambs, two yearling colts, and other little trinkets. Expenses were as large as I could stand, but I subscribed to a weekly newspaper called *The Universalist Terms*, at $2.50 per annum.

Potato bugs trimmed the leaves all off of our potatoes and the newspapers said that two women were poisoned by potato bugs in

Minnesota. The insect known as the Colorado potato bug had
made its first appearance in these parts in 1866. He probably had
been in other parts before, but to us he was a strange object and
many wild stories were told regarding his nature—poison, they
said, deadly poison.

When Mr. Bug first appeared of course, being a stranger, he
was an object of curiosity. His children were very numerous and
seemed entirely given to feasting, governed by nothing, for their
parents could not control them. They made sad havoc with our
potatoes for five years—the only way we could raze them was to
pick the buglets off by hand. But by 1871 they had almost disappeared and we had no clue as to why. It was supposed that birds
destroyed them, but why had these birds never eaten them before?
The bug still lived, though in vastly diminished numbers, and did
no major damage after that. So, after years of experience we
concluded the potato bug was a harmless, nasty thing and no more
poison than any other disgusting bug or worm.

Harvest of small grains was good in 1868, though wheat had
dropped to $1 and potatoes were not very plentiful, as the bugs
had raised hob with them.

Two of our neighbors had unfortunate experiences with their
horses that year. One of Daniel Thatcher's horses died when it fell
into a sinkhole while plowing. There had been no sign of a hole at
all in the field when, suddenly, his off-horse went down through
and the near one fell in onto the first. He succeeded in getting the
top one out by drawing him by the hind leg some twenty feet, but
the other one was dead.

In a separate incident, G. W. Horton's horses were stolen and
no trace of them could be found. They were missing from the field,
and we searched the woods on the ridge over the creek for hours
with no luck. Six months later, G. W. Horton accused Lyman
Howe of stealing his horses. Someone saw Howe in the vicinity of
where the horses were found, but he maintained his innocence,
saying he was in that area visiting his son's place. The horses must
have wandered up Willow Creek through the bottom land at
Thatcher's farm and the thickets around the Burnham place. They
could have lived undetected for some time in the uncleared areas,

or they got away from whoever took them. The horses were mangy and matted, with their manes and tails in knots and burrs around their hocks. As Horton had no thief to match them, he probably believed it was old Howe as long as he lived.

**

General Grant was nominated for president in 1868, but politics became a humbug. Our Congress had been at work since the first of March trying to impeach the president of the United States, Andrew Johnson. They squandered time and money without stint and acquitted the vilest man that ever wielded the scepter of authority in our land. Not only was it reported that he appeared intoxicated at the inauguration ceremony after Lincoln's assassination years before, but he pardoned many of his wealthy Confederate friends after the war. Then he insisted on battling Congress over reconstruction plans and twice dismissed the secretary of war. He was acquitted by one vote only, which voter would have served the country better had he been home ill that day.

Johnson lost the Democratic nomination, and Seymore of New York announced he was running against Grant for the presidency. The state of our nation was very critical at that time—if we were to believe the papers, we came very near having another war. Of course, the usual amount of lying and blackguard went on that attends a presidential campaign. It was time such lying and tomfoolery were done away with.

The third day of November was our national presidential election. I took a day away from my work to ride to Carimona, fourteen miles away, and cast my first vote for General Grant, who was elected by about two-thirds of the legal votes of the land.

This was when we first heard about some secret "Klan" in the South called the Klu Klux, that killed and maltreated unarmed Union men. The Union Pacific Railroad was finished that April, and we at last had a railroad stretching from ocean to ocean, from Maine to California.

— o —

Bottom of the Ladder

Like many young people starting out, when Fred married, he and his bride depended on their families for shelter. Before long, they set up their own home, yet stayed near Uncle Fowler's homestead.

My father, David N. Cummings, came to visit Minnesota from Vermont in 1868 and he liked it well enough to stay. Brother B. Frank Cummings bought himself a farm nearby, and I purchased some land near Uncle Fowler's property; forty acres at $5 per acre, with a first payment amounting to $38.

In October, Frank married Miss Janet Bowden. My childhood friend and schoolmate, Helen Powers, married Jim Moore on November 10th; and I, not to be outdone, married my best girl, Miss Rosannah M. Howe, on November 15th.

I first met my Rose in June—a perfect stranger. In July, I "scraped" acquaintance at their home. By chance, in passing one Sunday, singing arrested my footsteps and turned me in that direction. I heard some sweet notes and an angelic voice raised in hymn, and I had to find its source. This is all the excuse I shall ever give for our first meeting.

As might be expected, our wedding was no swell affair. No trip to Europe, only a ten-mile ride through November mud taken in a rickety lumber wagon. The ceremony was solemnized by a Methodist minister at the home of the bride's aunt; only that and nothing more. We both realized that we were poor and must work for a living, and save, if we would live in this world of selfishness and greed on the one hand and foolish waste and extravagance on the other. I came home the day following and went to work as usual and kept at it ever since, and the other party to the contract has done the same. How we succeeded in our efforts will appear later. Not without losses and many mistakes, to be sure, which

29

only emphasizes the fact that hard work is absolutely necessary to offset the blunders made in calculation and management of the seemingly trivial affairs of life.

Rose was taken sick with typhoid pneumonia in December, and doctor Case broke it up and charged me $12 for his services. Brother Frank's wife was sick with scarlet fever in early 1869 and there was much sickness in our neighborhood that winter. That was also the spring that smallpox broke out in Preston. Without any restraining quarantine laws, I was thoroughly frightened over the recklessness of those who had been exposed and incurred the enmity of some of them by some things I said. I thought they should keep to themselves and not go spreading dread disease over all the land. Many did not believe the theory printed in some papers that illness could be spread by breath or touch. There was more smallpox within a mile of us than I had ever heard of before. George Drury got it in Preston and gave it to his neighbors. The board of health determined that Valentine Pfremmer, who had recently come here from abroad, was infected and declared his house and all its inhabitants to be under their regulation. They dubbed the house a "hospital" and, for the purpose of public health, required everyone to stay away from it except for a few health officers.[*] While only one death was reported—a baby a few weeks old died—I feared for Rose's frail condition.

I worked four months for Uncle Fowler, receiving $11.90 that was paid to the state—the interest on my land debt. I got four acres broke on it after fifteen days of clearing brush, stones, and trees. I planned to build a log house on the clearing, and fenced part of it with posts and poles. I traded off my colts for a yoke of steers and $95, which showed that oxen were still used, and quite fashionable. I liked to drive them on the farm and sometimes think I should yet, except for the shock it would give my neighbors and the risk I might run of an insane inquiry.

Father David's troubles broke out afresh in March of 1869 when he received word that the woman he married to take my good mother's place had divorced him while he was out of

---

[*]   *Preston Republican*, March 26, 1869.

Vermont. I, myself, didn't weep because I loved peace and now hoped he would have it and, incidentally, give us a rest. This hope, I soon found, was not to be realized.

Rose convinced Father to stay in Minnesota, and he bought a farm nearby. A woman contrived a way to introduce her mother to David. My brothers and I did not learn of this until after it was too late and, though we did not care for the daughter or her brutish husband, we did not know the mother's character enough to speak either for or against a match.

Consequently, the Fourth of July 1869 was celebrated by Father in another matrimonial venture. Brother Moses also chose this date to embark on the matrimonial sea, and prepared to depart west for Cottonwood County, Minnesota. Sister Kate was married the 28th of July to Andrew VanSickle, which also would prove to be an ill-fated match.

The family expanded further as my little wife presented me with a fine boy September 3rd about one o'clock in the morning. We called our boy Warren Ellsworth Cummings. The baby was doing fine at the end of the month, but my wife was still not able to be up and I was discouraged about her health. Eventually she got better and I paid her sister, Mary Howe, $3 for house labor while Rose was sick.

I had cut some logs over the year and started to build a house. I bought the timber of Uncle Fowler: forty logs, sixteen rafters, eight sleepers, and seven beams. He charged me $17.65 for my house logs and $6.62 for some fencing I bought that summer. Not yet reckoned up were twenty-four square feet of studding, one hundred square feet of siding and four square feet of sheeting, which was still to be added to my home. I settled with Uncle Fowler and found $20.61 my due after reconciling my share of the harvest and deducting $50 for board. I brought Rose back to Uncle Fowler's from town where she had been staying with her parents, and we settled in to winter chores.

At the end of 1869 we had a little boy, almost four months old, and I couldn't find much fault with the dealings of providence. We called it hard times, I suppose because wheat sold at 50¢ per bushel and our farmers depended on wheat to pay their debts. As for myself, if it wasn't just as it was with me I should be hard up, but—by careful management—I tried to keep my head above water.

Over the next few months, I fixed my house so that it kept me warm, but it was so far from completion that I didn't dare guess when that much-looked-for hour would arrive. Even so, we moved in the last day of March, 1870, and commenced housekeeping with things as follows: a $33 stove; a $5.50 table; a homemade bedstead upon which were two good new quilts that Rose made, three sheets and a straw tick, with pillows, of course; a boot-box for a cupboard; three chairs; and household goods amounting to a washtub and board, plates, cups and saucers, cutlery, chairs, table cloth, salt dish and pepper shaker, spoons, and enough more such stuff too numerous to mention.

The work I did that spring amounted to planting twenty-three acres of wheat on the old homestead, three acres of corn, half an acre of garden and potatoes, and one acre of beans. My corn was the best that there could be, being eighteen inches high by the end of June. I had the following work on hand: corn and beans to hoe, five acres of grubbing, thirty rods of fencing to put up, a cellar to dig, and a house to fix.

Cattle were dying in this part of the world that year. In September I lost my pet cow, the first one Uncle Fowler had given me as a calf, years ago, in exchange for farm work. One of my oxen was sick, too. Uncle Fowler lost five and more of his were sick. We hoped that the cattle plague was stayed after a month, but then the rest of my cattle got sick. I panicked and sold them for a $75 note, payable later. I then bought a pair of steers for $45. Uncle Fowler buried eight cows in all. No one knew what was making the cows sick, or why some died and others lived.

We were visited by the cattle plague again the next year, but only one of our herd got sick. By then most of us believed it to be caused by toadstools which sprang up during the wet weather and were eaten by cattle allowed to run in the woods. No cattle were sick but those that ran where the things grew, and they were known to be poison to men.

That irresponsible who bought my cows did not pay me, and I was sick at thinking I had been duped by someone I trusted. He gave me excuse after excuse. I had my patience so tried with the man that three years later I finally sued him for the money he owed. I got a settlement then, but the whole experience was so distasteful I resolved not to sue another man. How long I kept the resolution may appear later, though I may as well own that it took one more lesson of the kind to sicken me completely of appealing to the law for redress. I finally accepted thirty-three and one-half bushels of wheat as the last installment on the debt for those cattle. I was constrained by this episode to make this resolution: all my future dealings were to be in cash or goods. I lost much peace and some money by a system of trust which gave my debtor the soft side of time while I took the hard side for my security.

My brothers had been hired to help me clear my land that year, but disappointed me about the breaking and let me have a cow as part of the reconciliation. This helped me build my little herd back after letting those six go. I also hired Father that harvest, paying him two bushels of wheat per day.

There was a funeral at Waukokee as Rose's sister, Samantha Gould, buried her baby early in December. News from my brother Moses, who had taken up residence out west in Cottonwood County, informed us that he had a son.

In looking over the year, I found myself just about where I was a year before. I had spent all I earned, and more. My luck was what folks called bad. I lost some money on those cows, and on hiring work done on the farm. But I had seventy-five bushels of wheat, three pounds of hay seed, and straw enough to winter my stock, which I thought would about balance the account. I hired on to teach school again the winter of 1870–71 in the same district in which I had labored for three years. I got $25 per month for three months, besides the hospitality of one Mrs. Bowden. With revenue thus obtained, I kept the wolf from the door.

We lived in our own log cabin on the claim and worked the old home place, making it an extra-hard job, and I have to say a homesick one. Although I loved my new home, I could not drive away the miserable longing for something hard to define. We began housekeeping at the bottom of the ladder in truth and in deed.

— o —

War in Europe

Fred followed current events through several newspaper subscriptions. The following section summarizes his understanding of the Franco–Prussian war, and demonstrates his horror at man's inhumanity.

The news in the fall of 1870 reported a war between France and Prussia. Their history, as near as I knew, was this: France declared war against Prussia and invaded it. Prussia took up arms, drove the French back, invaded France, captured Emperor Louis Napoleon III, and got within eight miles of Paris, the capital of France. After the capture of Louis, the French declared France a Republic and much fighting and slaughter followed. It appeared that Paris was besieged.

The bombardment of Paris commenced in December, and the siege continued on steadily. The Prussian army was reinforced and hope for the French was dying out in this country, for it looked as though France must submit.

The French people's suffering was very great. Hundreds starved and many resorted to eating dogs, cats, and even rats, but still they resisted the march of the victorious Prussians and contended foot by foot the ground which they could have no hope of holding. After a siege of about eight weeks, the French were compelled to surrender Paris in January 1871. The Emperor of Prussia proclaimed an armistice and sent supplies to the suffering French, which we took as evidence of true manhood, and corresponded with the true spirit of the brave warrior.

The prospects for peace improved early in the year. The French came to their senses a little, but it was evident that the much-talked of Republic was played out. Probably, we thought, Louis would return to the head of government, although others contended for the place. Peace between the Prussians and the French was at last concluded in March, the conditions of which restored France to her nationality while the Prussians received territory and a payment.

Then the French fought among themselves about what kind of government they should have and who should rule them. There was much bloodshed in the city of Paris, with at least one hard battle in which one side was reported to have lost 2,000 men. The prisoners taken were shot, it was said. The papers in May still reported on the sad havoc in France. Paris was again besieged and bombarded. Men, women, and children were exposed to the shot and shell of those professing to be their friends.

The French war finally closed in June 1871. It was a bloody time and ended in a most inhuman massacre. The city of Paris, after a heavy siege, was taken by storm and its defenders slain by the thousands, and many of the wounded were buried alive. The sight was said to be horrid beyond description—15,000 men and officers were said to have been butchered. What kind of a government would grow out of it remained to be seen. It was such a mixed concern, I hardly knew what they were fighting for, except it be power.

—o—

Lyceum

Before the advent of radio or television, people had to create their own entertainment and venues for intellectual stimulation. Leisure time was very limited and activities had to be worked around a farmer's busy schedule. Fred participated in various community gatherings, but also spent part of each day reading, noting events in his journal, or writing poems.

In January of 1871 several of our neighbors formed "The Waukokee Vigilance Association." James Hipes, Martin Kingsbury, William Wilber, William Tinkelpaugh and others banded together against horse thieves, some of whom had their abode in Fillmore County. A few weeks before, a known horse thief had come to Preston. A warrant for his arrest was drawn up, but put in the hands of a drunken officer who was not in a condition to discharge his duty. This gave the villain time to escape. The next night, William Wilber's horses were stolen. Many thought the vigilance association would be more effective than the law at dealing with these thieves.[*]

For literary exercise, us boys got together in February at the schoolhouse in Waukokee and organized a Lyceum. We pledged to meet every Saturday evening to debate current issues. We numbered sixteen members. The first week we had a lively discussion on "The Miseries of War and Intemperance." The judges decided that, according to the arguments, war had caused more misery than liquor—but intemperance was restricted to the immoderate use of liquor. The next week the members of the Waukokee Lyceum met in our room and, after a very amusing discussion, our judges decided that women had not the same

---

[*] *Fillmore County Republican*, January 27, 1871.

natural right to the ballot as man. In March, after another lively debate of two hours' duration, our judges decided that the Indian had more cause for complaint than the Negro of the treatment which they received from the white man, according to the several arguments. In looking back, I think this was due to the fact that in Minnesota we had more direct experience with treatment of the Indian, and am not sure that the argument would go the same way if it were discussed again. We debated the question of war and intemperance again and, being more evenly divided than before, the affirmative won the question according to the judgement of two out of three men.

Our Lyceum prospered finely for a time but a lingering memory of it, like many other recollections, gives me a tired feeling. At one meeting I recited one of my poems, "My Dream."

My Dream

I dreamed a dream, a vision rare,
'Twas caused, no doubt, as others are
By too late supping on too sumptuous fare.
I thought my days were lengthened into years
And that I trembled on the verge which every mortal fears.

Then striding proudly through the sky
The conquering specter, Death, drew nigh
Alas, my time had come to die.
The clammy deathdamp gathered on my brow
And through my ridged veins the blood ceased to flow.

Near Death, whom man so long has feared
As a grim monster, conscience-seared,
Another being now appeared;
Kindness alone moved him to strike the blow
That freed my fettered soul and bade my spirit go.

Then 'twas new sight given to me
Earth's mists and darkness fled away
And heavenly visions could I see,

For friends that I had thought would speak no more
Now gathered round to greet me as in days of yore.

My little clock with nervous stroke
Chimed out the hour and I awoke,
But before the rapturous spell was broke
Such soul-inspiring music fell upon mine ear
As never living mortal had permit to hear.

O man, thy life is but a dream
Where rank extreme meets its extreme,
And hope and fancy reign supreme
Nothing on earth is altogether sure;
Living we die and dying live, more real than before.

And when I finished reciting it, one of our members called out, "Mr. President, I move that we don't have any more such stuff and nonsense allowed here." This foolish speech of mine leaves a bitter taste. I shared some personal thoughts and philosophies, tried to get a spiritual debate going, and was ridiculed for it.

— o —

"Who Will Miss You When You're Dead?"

(My First Angel Visit Recorded)

One night while I dreaming lay,
Resting from a toilsome day,
Came a vision to my sight
Of an Angel clothed in white.
And to me the Spirit said:
"Who will miss you when you're dead?
Of your care, your labor here,
None will know or care to hear."

"Here you labor ceaseless, quite,
To instruct your child aright—
Clothe him, feed him day by day,
Guide his feet in wisdom's way.
But he'll forget you when you're dead,
When the grass grows o'er your head
Other thoughts will fill his mind,
He'll have better grists to grind."

"Your lovely wife whom you caress,
Soothe and comfort in distress,
When she finds her 'first love' dead
Soon another man will wed.
She'll not miss you when you're dead,
When the first great grief has fled
Other's love will dry her tears,
Another voice will charm her ears."

"Gentle Angel," I began,
"Prithee bear with mortal man
And I will shortly tell thee why
I'm content to work and die.
Although none miss me when I'm dead
And my labor shall have fled,
I'm fulfilling God's command
Plowing up the fallow land."

"God, my maker, placed me here
Bid me labor in his fear
And on him I place my trust
When I give my dust to dust."
Then methought a quiet grace
Shone upon that sinless face,
From her lips this sentence fell:
"Mortal, thou hast answered well."

—1870

School Management

I worked my claim and for the neighbors: for William Tinkel-paugh, setting out apple trees and for Martin Kingsbury, corn planting and cultivating. I bought a dozen apple trees of Tinkel-paugh for $4 to be paid for in work. I worked to improve the farm that year, and this, with supporting my little family, kept me in business.

I had the bad luck to lose my cow, and struck a deal with Martin Kingsbury to buy one of his. The price of the cow was $30, to be paid for in harvest labor. She seemed to be a very good cow, and she had a fine he-calf that spring.

I had intended to pay for that cow I bought in harvest work during the fall, which I failed to do. So I requested Kingsbury to take it back offering to pay for the use of the animal, and felt a sense of relief when he kindly took the cow and calf and released the obligation. I gave him the calf and two days' work as a recom-pense.

I sold Rose's brother six cords of wood for a violin, the wood to be taken in the tree and grubbed up, and traded a watch and $5 for an American Shuttle sewing machine. Five months later, I traded that sewing machine and violin for a wagon, which was priced at $45. It was time to be practical, and we needed a wagon much more than those other luxuries. I worked for John Vail in the harvest during which time I earned $32, as we needed the money.

I prepared for teaching again in the fall of 1871. We had a teacher's institute at Lanesboro, which I attended, working so hard as to bring on a congestion of the brain that nearly put me out of business for good. Dr. Ross and my good wife saved the little life I had and again enabled me to take up its labors and disappoint-

43

ments. Hints for school management and instruction received at the institute:

- Everyone must work for himself.
- Throw the child overboard when you are sure he can swim.
- Never allow bad grammar in school.
- Teach scholars to talk proper when at play.
- Always require pupils to talk plain and make a full sentence when reciting.
- Lay the foundation of grammar in reading.
- Require pupils to tell the elements of the sentence read.
- Select words that are mispronounced and write them.
- Require words pronounced until the pupils are masters of them.
- Teach spelling in every branch of study.

My work had been hindered very much by my sickness, during which time my corn was nearly destroyed by cattle as we had no fence.

The fall of that year was remarkably dry, and Chicago burned to the water's edge. This was the most destructive fire the world ever saw. The city, it seems, had been built almost all of wood structures, which was dry as kindling with the heat and lack of rain that year. The devastation burned 18,000 buildings over more than 2,500 acres. It was said that 100,000 people were homeless and 120 lives were lost. One could hardly imagine the scene of such ruin; black, charred posts and rubble that once were stately homes and business quarters. I was sick at the time, and worried and tormented by the reports brought in by indiscreet visitors who would probably be invited out of a sickroom today.

Trials and perplexities so increased and multiplied around me that I concluded to sell out and move to the west, where brother Frank had settled and Moses was going in the spring. However, money matters would not permit it as I couldn't sell the farm right then. Additionally, the winter had turned very cold and many people froze on the western prairie, which warned me to stay here until I got money to build with and so, between poverty and fear, I now expected to remain.

Therefore, I returned to work again at my old game, teaching. I was hired to teach three months at $25 per month, plus board, in the town of Bristol Grove. Rose and my son were living with her sister and brother-in-law, the Goulds, that winter.

District No. 151 was a mixed multitude, half Welsh and half Norwegian. They had a new frame schoolhouse, twenty by thirty feet with room for fifty scholars. It was a solid building made of logs with the cracks between filled with split sticks and plastered with lime and sand mortar. The desk boards were mounted on the wall at an angle. On the top of the desk board was a four-inch strip nailed on the level, to hold the ink bottles. The pupils, when engaged in writing, had their faces turned towards the wall, and when not engaged in writing they faced the center of the room using the edge of the desk-board as a rest to their backs. My school was rather backward, and most of the scholars were young and small.

That year I wrote several poetic compositions. The first, titled "Reflections on the Dying Year," was blocked out at my boarding place in Bristol on December 31st, 1871. "Minnesota" I guess was also written in 1871.

Then, I quit my school twelve days before the contract stated, rather than continue a hopeless fight in which I had nothing to gain and everything to lose. My Waterloo, perhaps, as my enemies were unscrupulous and treacherous, and I, harmless as a dove, had not the serpent's wisdom. The historian tells us that the great Napoleon's fate was sealed by a passing thunderstorm on the night before the battle and a sunken road upon the battlefield into which his cavalry plunged. And circumstances as insignificant as a few drops of water more or less started a train of unfortunate "scraps" that finally drove me out of Bristol Grove.

Of course, I might have stayed and drawn pay for the remaining twelve days, as most unwelcome teachers do. But I just packed up and started home the same day as I told them, "While I stay in this house I insist on having order." And when I could not have it I would not take pay, nor waste time with them. My pupils were an unruly bunch, and their parents would not support my disciplinary measures. The older children picked on the youngest, and sassed at me when I tried to correct them. My observation tells me that a teacher must have the good will of his pupils, and I can't stay where I am not wanted for love or money.

As the year drew near its close, the day of reckoning showed my accounts stood thus: money received, $125.03; money expended, $125.86. My debts amounted to $30; my credit was ditto. My luck that year had been bad but, notwithstanding, my heart was light with hope of future prosperity.

−o−

Reflections on the Dying Year

I am sitting tonight 'neath a friendly roof
Watching the old year die,
And casting a glance at the dim, distant past,
At the years, months and days gone by.
I am thinking tonight of the years that have passed
Since this poor life began,
Of my crooked paths and my many sins
Which like mountains 'round me stand.
I am thinking tonight how vain my attempt
To rid this poor soul of its care,
And when I look to my neighbors for help
I find that they all have their share.
I am thinking tonight of that better land,
The home where the glorified dwell;
But oh, will it ever be mine to enjoy
And hear the glad shout "All is well!"?
Oh Father in Heaven, hear thou my prayer,
While watching the old year die,
Oh save me in thy mercy from
This pit of miry clay.
I ask not for riches, glory or power,
Which pass like the vapor away,
But oh, give me meekness, courage and faith
To meet life's duties each day.

—1871

Minnesota

I have read of the pines that in Oregon grow
And their vast mines of wealth 'neath perpetual snow,
Of a fair, southern clime where the birds ever stay
With no chill winter's wind to drive them away.

And my soul with a patriot's pride contemplates
The glory and wealth of this Union of States;
Yet of all our possessions in East, South, or West,
My home, Minnesota, seems dearest and best.

Her waters are pure as the river of life,
And her fields with the glories of Eden are ripe;
The garden of God never knew purer air,
Nor Sharon can boast of a flower more fair.

The Indian wept and in grief broke his bow
When from fair Minnesota he found he must go;
So I, too, should sorrow in deepest dismay
If some dread misfortune should drive me away.

'Tis the home of my youth and I love it full well
And could e'en be content here forever to dwell,
Now a word to the wealthy, the poor, and distressed,
Minnesota invites you to come and be blessed.

—1871

Soreheads

Fred followed each national election conscientiously. He felt it was every man's duty to vote and always took the time needed from his farm chores to travel 14 miles by horse and wagon to cast his ballot. He had strong opinions about the candidates and campaigns, which he monitored through newspaper reports and an occasional speech by local delegates. Fred was a staunch prohibitionist his entire life and he supported the women's right to vote movement.

Papers in early 1872 brought accounts of earthquakes in diverse places. There was one in California which threw down houses, spreading destruction in its path. Another earthquake occurred in Syria, which killed 1600 persons. Mt. Vesuvius, in southern Italy, again became a terror to the people who lived in its vicinity, and towns within its reach were buried in lava.

The political war was getting warm and excitement began to rise, as 1872 was to be our presidential struggle. Prospects suggested that the country would be divided into three parties, to wit: Republican, Democrat, and 'Soreheads,' which latter cried "anything to beat Grant." In a message to Congress in April, President Grant introduced the rights of foreign immigrants in which he recommended legislation that would secure them safety and comfort on shipboard, and prevent rogues from imposing on them and getting their money by fraud. He said, further, that efforts had been made for two years to secure the aid of other nations in this effort with fair prospects of success.

As predicted, the Philadelphia convention unanimously endorsed Ulysses S. Grant and raised his name to the masthead for a second time. Horace Greeley was nominated in Baltimore by the Democratic Convention. The field was now clear to Grant and Greeley—an obstinate general and a fickle-minded editor, the Fanner and the Farmer. There was another campaign convention called the "Unpurchased Democracy," which nominated Charles O'Connor and J. Q. Adams, but the gentlemen declined the position.

The following was the last plank in the Republican Platform: "We believe that the modest patriotism, the earnest purpose, sound judgement, practical wisdom, incorruptible integrity, and illustrious services of Ulysses S. Grant have commended him to the ballots of the American people, and with him at our head we start today on a new march to victory." They also had a woman's plank which, after acknowledging the high service rendered the country by them, promised to give their claim for additional rights due consideration.

In looking over the record of 1868 and comparing it with 1872, I was unable to decide which was the worst political campaign. Every kind of lying imaginable was going on: Greeley was called "a turncoat, a liar, a Judas, a Klu Klux, and a sorehead." Grant was "a miser, a dummy, a fool, a harper, and a thief." All this stuff no one believed but, if they did, it looked to me as if the minority in either case would be justified in rebellion. Mr. Grant had, for three years, faithfully served as chief of our nation and it was customary with us to re-elect a faithful man. In the past there had been at least five men re-elected to the presidency, of which Washington was the first.

I was very sure that only a victory for Grant would save the country, and voted accordingly. Also, laboring under the mistaken notion that a vote for Grant was one for prohibition and woman's suffrage, I took a double interest in voting the Republican ticket. When election day passed, Ulysses Grant was elected by a large majority.

Horace Greeley died that December; the veteran, editor, and statesman was no more. As near as could be told, overstrain on the mind was the cause of his death, as it was said he could not sleep after his defeat. Mr. Greeley's life was one of study and observation; as an historian, he had few superiors. For years he advocated the abolition of slavery, but his course for the previous eight years had not reflected credit on him, unless bailing murderers and going back on former professions may be called a credit to him.

— o —

Horse Plague

As more people he knew moved to settle new land further west, Fred entertained thoughts of moving. Torn between the ties to his boyhood home and an intense interest in exploring the world, Fred's need for stability after the insecurities felt during his childhood won out. He stayed put, tied to the land he loved, and contented himself by exchanging letters with family members in other parts of the country and by reading about exotic lands in newspapers and books.

The events of 1872 began by my disposing of my land claim in February with the determination fixed of going to the west. I traded off my forty acres and steers for a span of five-year-old mules and $25.

Uncle Fowler put a stop to my moving west to Cottonwood County by an offer that suited me, and which Rose was well-pleased to accept. My contract with C. P. Fowler was as follows: I would work his farm thereafter for a certain share, according to the amount of labor done. I may keep a cow, raise my own pork and a few sheep, not to exceed ten. So we came back to the old homestead and settled down to the business of farming and homebuilding.

Well, time and fortune found me once more at the home of my youth and childhood. What my future may be, time alone would tell. So far as fortune was concerned, I lived and kept myself clear from debt. I rented another five acres of corn ground with an unwritten contract as follows: the owner would provide seed and tools, and I would do the work and would get half of the yield to be taken standing. I also took a contract to haul fifty bushels of sand at 10¢ per bushel.

News from Vermont announced the death of my uncle Asa and aunt Harriet Cummings in March. That winter also carried away my wife's uncle, D. C. Phillips. It was quite sickly around our area, the prevailing diseases being scarlet and lung fevers. The daughter of neighbor Freemire was buried, and Dr. Ross called the disease that caused her death spotted fever. Everything that Ross could do did her no good, and Ross was our best physician.

In April, two more neighbors sold out their interest in Waukokee and moved to the west. We received word of their arrival in Cottonwood County after eighteen days on the road.

That spring was very windy. We also had what may be called "the flood" in early May. Willow Creek overran its banks very suddenly, rising higher than ever before, sweeping everything in its way. No lives were lost, but the mystery was, what caused the freshet?

I killed a wildcat that summer, which neither dog nor man could frighten from her young. By early August we were in the midst of harvest. The yellow shocks were to be seen on the distant hills in every direction. The clatter of the reaper and the dull chant of the harvest fly— always heard together—now called forth the harvester by day and lulled him to quiet rest by night. Wheat was in the shock or the stack by late August, and threshing commenced. Harvest wages were $3 here, although $4 was offered in some localities for a day's work. Taken all through from seeding to harvest, I believe it required as much patience to carry on a farm as any other occupation. Moses bought my mules and wagon and I took his note, endorsed by J. W. Green, for $250 at ten per cent interest.

I bought a house from Andrew VanSickle for $50 to use the materials in building my own place, and began the hard work of tearing it apart. We finally succeeded in adjusting the roof of my new house. It was a difficult job, but it was at last "raised" by seven men, with Father, my brothers, and a neighbor coming to help. My house was far enough completed that we moved into it at the end of September. This little house, framed by my own hands, was rough and "uncanny." But we lived in it a quarter of a century, and eventually left it with a sort of homesick dread for a better and bigger one.

A new horse epidemic broke out in the eastern and middle states which was spreading rapidly; it was called "epizootic," whatever that means. Oxen had to be used in the large cities in place of horses. The nature of the disease seemed quite similar to so-called distemper, and it swept the whole county clean. Hardly a horse escaped and most all suffered from its effects for a long time and, as usual in such cases, every man had his own peculiar notion of how to treat it. Some would not allow anything cold to be given, others gave cold water with snow in it. As for myself, I let nature have her way, fed light, and gave water, as it was their throats were sore and they would neither eat nor drink much at a time, so I gave a little and often.

Whew, how the wind blew that fall! We had a visitor come to our house November 13th and, although an entire stranger, he refused to leave. I suppose it was because it was so cold. As this baby was still with us, we proposed to call him Lewis S. G. Cummings. The little boy slept in our bed and increased the number eating around our table to four. Winter set in as the wind blew around the house like a lion—our house was not overly warm and I had my hands full to keep comfortable. I paid $6.25 to father Howe's family for housework while Rose was indisposed.

The horse plague came on apace in December—several cases were reported in Preston, in Rushford, and in Lanesboro. There was only one way to treat it, and that was to keep the animal warm—if he got cold it meant death to the horse. The eastern cities were entirely stripped of the convenience of horseflesh. We expected our horses to be taken sick at any day, and very few well horses were to be found. Traveling with horses had almost entirely stopped, and the disease was not confined to horseflesh entirely; fowls of every kind, deer, swine and cattle were also subject to the contagion. That one animal took it from another was very probable, but the rapidity with which it advanced and the fact that every horse was affected showed a general contagion without contact of any kind. Eventually, the horse plague subsided with comparatively few animals in our area dying from it, but still fewer altogether escaping its ravages.

—o—

Farewell to the Old Homestead

(When about to move west)

Farewell, my early home, farewell;
With grief I break the magic spell.
A halo here attracts my soul
As magnet trembles to its pole.
Ye fertile fields where I have toiled
And Heaven blessed an orphan child,
Anguish the rambler never knew
Wrings from my soul this last adieu.

Oft have I bathed my throbbing brow
In this sweet stream as I do now,
But ah, the application's vain—
The thought but aggravates the pain.
Wretched, I wander here alone
Among these shrubs my hands have sown,
A homeless fugitive I rove
A stranger in my native grove.

Yet there is hope, I'll not complain,
While youthful strength and health remain.
Upon the prairie broad and free
These hands can build a home for me—
A home where age may find repose
Where life may reach its final close.

—1872

Insult to Our Flag

We heard considerable stirring news in 1873 consisting of earthquake, murder, and startling revelations of fraud and treachery in high places. The city of San Salvador in Central America was destroyed by an earthquake, destroying 800 lives and $12 million in property.

The 42nd Congress adjourned *sine die* in March after covering themselves with shame unprecedented in history, if we may except the case of Andrew Johnson. They finished their session by appropriating $5,000 to each member of the "ring." This was called legislation by some, but others called it public robbery.

The farming public were organizing to make an attempt to break the chains of oppression which railroad corporations, wheat rings, and a treacherous lawmaking power were stealthily binding upon them. There was a new organization in the United States known as the Patrons of Husbandry, which at the beginning of the year was a mere trifle but by the end counted its members by the thousands.

General Canby, Brigadier General of the United States Army commanding troops in Oregon, was treacherously murdered by Indians to whom he had been sent without arms, as a peace commissioner. Orders were issued immediately for the extermination of the Modocs. But the Modoc Indians were not so easily conquered and it was feared they would cause much trouble and bloodshed. They had already killed many of our troops, about fifty, including a captain, a lieutenant, and other officers. Eventually, some of the Modocs surrendered voluntarily and unconditionally. Their leader, Captain Jack, and his remaining twenty men were reported to have joined another tribe. It was also reported that 5,000 warriors were ready for the warpath on the upper Missouri. Captain Jack, driven to the wall by the forces of General

Davis, surrendered after a short fight in early June. General Davis wanted to make short work of Jack and justice, but orders from Washington rendered the gallows that he erected another monument of lenity to the ever-merciful ruler of America, U. S. Grant.

A most destructive tornado swept over Iowa toward the end of May. Houses, barns, fences, and everything movable was caught up and whirled through the air like toys and then dashed to pieces. Many people lost all their property and many were killed. The tornado was about a half-mile wide and spent its force at Cairo.

Our government made a demand of Spain that fall for an alleged insult to our flag and the barbarous murder of a ship's crew. For the last five years, Spain had been murdering and robbing the inhabitants of Cuba, and it was one continued scene of blood. The aforesaid vessel was accused by them of carrying aid and comfort to the enemy. It was thought there might be trouble yet with those butchers, and our navy was being rapidly made ready for war.

Spain refused to make amends for the massacre on *The Virginia*. General Phil Sheridan was ordered to prepare for action, and things began to look squally. President Grant's demands were: first, an apology from the Spanish government for an insult to our flag; second, Spain was required to provide for those whom they had rendered widows and fatherless; and third, those who still lived in prison must be released and recompensed for false imprisonment.

A short time later, the president's message came to hand in which he said our trouble with Spain had been settled to his satisfaction. Mr. Grant also asked Congress to pass an act which was clear evidence that he disapproved of the salary steal of the last session. The act in question was to the effect that the president should have power to veto a part of a bill without disapproving the whole, also that Congress shall not pass an act within twenty-four hours of its adjournment.

War and bloodshed continued on all over the world throughout the fall, although our government succeeded in settling all its difficulties and lived at peace with other nations. Spain was in a state of war, both at home and abroad, with her own subjects. Russia was invaded and warring against India. How these wars were going did not appear in the newspapers, except that lives and property were being wantonly sacrificed. The papers also said that the notorious Captain Jack was to be hanged October 3rd with several others of his tribe.

— o —

Rapid Multiplication

January 1873 produced vast quantities of wind and snow, which piled the snow into the lanes to such an extent as to stop travel entirely. Willow Creek froze up to the bottom and ran truant all over the valley, freezing as it went, making a fast sheet of ice. Further accounts of the storm told of dreadful suffering and death.

A family froze to death four miles from Granger, under horrible circumstances and this was but one instance out of many. Mr. Evans, a Welsh minister was returning home with his wife and two children when the storm overtook them. A snowdrift blocked their way within a mile of his own house, so Mr. Evans carried his oldest son home and returned to the cutter with some extra blankets for his wife. He then started for the house with the baby, but must have lost his way in the violent storm and perished. Several days later, when neighbors called at the house they found a little frozen occupant inside and, down the road, the stiffened corpses of the horse and the wife.[*]

Many cattle also perished in the snow. Men who had lived in Minnesota for twenty years said this cold had never been equalled. The loss of life and property was immense, but owing to the almost impassable blockade of snow, no definite news could be received. We finally heard from our folks in Cottonwood County that the pioneers were all well. None froze up during that terrible storm, but they would be very apt to remember it for some time.

By March we began to talk of winter as a thing of the past. The heavy load of snow and ice that covered Mother Earth like the pall of death left us for other places. I heard a bluebird on the morning of March 16th. A most incredible account of cold weather

---

[*]   *Preston Republican,* January 17, 1873.

appeared in the paper that told of suffering in Nebraska almost equal to that caused by our Minnesota storm.

I heard that an old schoolmate of mine, F. B. Drake, who was practicing medicine in Chicago, Illinois, died of smallpox. He was a young man of rare talent and sterling integrity.

> They say he has gone to the regions of death,
> That the pitiless grave encircles him 'round.
> Yet still I have hope, though he's yielded his breath,
> That his spirit can never be bound.

Several of our friends who had moved west returned to this vicinity to help with the harvest. They reported Cottonwood County was badly devastated by grasshoppers and nothing was left to pick in the fields as the locusts had stripped them clean. They needed work, and money to send to their families, and harvest labor that year earned $3.50 per diem in these parts. Our harvest looked very promising. As far as I was able to ascertain the crop would be extraordinary, and such fine weather to gather it! Farmers here were highly favored that year.

> A glorious harvest is with us again,
> 'Tis a time of rejoicing and cheer,
> The hilltops are crowned with the rich, yellow grain;
> In the valleys its beauties appear.

The busy rattle of the sickle ceased by the end of August, and the yellow stacks loomed up away on the distant hills. The weather was very dry all through harvest, and wheat was a number one crop. We put up eight stacks of grain, five of which were wheat. Corn was already ripe by early September.

Our good fortune with the harvest was to naught when there was a panic among wheat speculators so that wheat prices went very low and it was hard to sell, in fact, there was no market for wheat at all. The panic was caused by the breaking of a number of wealthy banks, the principal of which was owned by Jay McCook, who was said to have lost $4 million in railroad stock, all caused by gambling.

Our babies were sick with diarrhea and *cholera infantum* that year. The weather was lowery with now and then a shower of rain. During that autumn the South was severely afflicted with yellow fever. Thousands fell victim to this terrible scourge. Cholera also committed fearful ravages along the Mississippi River.

About this time my mind was exercised on the subject of spiritualism. I had seen manifestations that, although they did not convince me of a real supernatural agency, kept my mind agitated and curious to pry into the issue of spirit power. Spiritualism got a new twist when Captain Hoyt Winslow gave or, rather, sold lectures in Preston. I took them in and thought them grand, but it took the charm off when, a few weeks later, I was told by one of his admirers that he was in a mad house.

Then there was a sudden change in my mind that turned my thoughts from things spiritual to things natural, caused by the sudden appearance in our family of another little blessing dated November 17th, we called him Reuben A. Cummings, and he was the light of our lives.

Lines to the Baby Boy

Little pilgrim on the border
Of life's rough uncertain sea,
Do not think it out of order
Deign to answer if I ask a thing of thee.

What can be thy mission, stranger,
In this busy world and wide?
Art thou Mason, Jew or Granger,
Tell us, stranger, where thy latent powers hide?

Here are hands unfit for working,
Far too small to hold the plow.
Clumsy, nervous, sprawling, jerking,
Tell us, stranger, what these little hands will do.

Little toes with feet to match them
Set on legs ill-shaped and weak,

Flesh and cords in perfect stratum;
How I wonder what this little wad bespeaks.

Ne'er a tooth can I discover,
Lips and tongue no office fill,
But thy voice is strong and clever
And I tremble, lest things go against thy will.

Altogether, little baby
You're a wonder in the land,
Speak, and tell us maybe,
What's the labor for thy hand?

As was my custom at the close of each year, I again looked over
and reviewed the important events of 1873. My personal affairs
stood about square with the world. My accounts showed $92.50
received and $93.50 expended, and I had about sixty bushels of
wheat remaining. I owned five sheep, a cow, calf, and pig. I lost
two sheep during the year. My brother failed to meet his note due
last September, so I still held notes amounting to $225, and my
total indebtedness amounted to about $17.

I wrote several quite pretty poems this year. That is, I think so.
Our house was quite a pleasant one and we lived in hopes of
having it better. A third beautiful boy baby came to gladden our
hearts and my family amounted to five persons at the close of 1873,
which I counted as rather rapid multiplication. I had three boys:
Warren, Lewis, and Reuben. My little poem, "Watching the Fire,"
may be of no great literary merit, but it gives an idea of an interesting characteristic of my youngest boy: that of a seeming media-
tion, as though studying some deep problem. My imagination,
doubtless, but who can dispute the workings of a baby brain?

Watching the Fire

Perched upon papa's knee,
Snuggled so cozy and warm,
As thoughtful and sober as baby can be,
Nestles a beautiful form.

Tired of mischief quite,
Does his little soul ever reach higher?
On his face is a heavenly light
And he looks like an angel tonight,
Sitting here, watching the fire.

— o —

Funeral in the Woods

(Composed for October)

There's a doleful sound in the woods, 'Zanna,*
Where so lately 'twas ringing with glee;
The chill, moaning winds and the rustling leaves
Make it seem like a funeral to me.

The maple and poplar are bare, 'Zanna,
Where the songsters in June were so gay;
The keen, cutting frost and the rough, western wind
Have driven their verdure away.

The robin and swallow have flown, 'Zanna,
From the winter so frosty and gray;
And the bluebird is chanting the last plaintive note
Of her dirge in the forest today.

She tells of a time that's to come, 'Zanna,
When the frost of age and of care
Shall dim the eyes now bright with hope,
And whiten our smooth, glossy hair.

There is trouble and sorrow and death, 'Zanna,
Ever lurking to make us their prey,
With a blighting grief or some sickening woe
When they seem to be farthest away.

But the springtime is coming again, 'Zanna,
When green will be seen in the grove,
And the birds will return with their rich, merry songs
That speak of contentment and love.

They will sing of a future for us, 'Zanna,
When we've passed through affliction and pain,
Where the waves of despair or the breakers of woe
Can never assail us again.

—1873

---

*   'Zanna is a pet name for Rosannah.

Domestic Warfare

The con man described in the first story here might not seem too successful by today's standards, but considering that $100 represents one family's income for an entire year, the loss was hard to bear and made it noteworthy. The family had other troubles, too, though Fred was usually very discreet about recording such things. However, the drama surrounding his father's third marriage was too big to ignore or gloss over. Fred not only comments on some of the circumstances in his journal, but also pens a short fable using archaic language that clearly illustrates his disgust with the situation.

It was in the winter of 1874 that a swindle was recorded. It was pulled off on the Colburn family and so cleverly planned as to deceive Nathan Colburn and Uncle Fowler. The story began about eighteen years ago when a young man named Robert married Uncle Fowler's niece, Elizabeth Fowler. Both were living in Minnesota where Robert took a claim with a lively hope of making a home among the "wilds of the west," for such it was then called. But his hopes were doomed to disappointment. Lizzy died before they had been married three years, leaving a little baby some six months old or so. Robert then left the baby, Florence, with Lizzy's aunt and uncle, Ruth and Nathan Colburn, rented his farm out, and went to work elsewhere but was soon taken sick. In the winter of 1861 he went back to Vermont having made arrangements with Mr. Colburn to care for Florence until his return. But he never came back, as he died about two weeks after he got home. In due time his farm was disposed of by the probate court, with the proceeds appropriated for the support of the child and other expenses.

Years later, in 1874, a man claiming to be Robert's brother came here declaring the aforesaid sale to be null and void and promised to restore the estate to Florence or sink his fortune in the attempt, which fortune he intimated to people was heavy. He commenced his job in style by hiring a lawyer to overhaul the papers and look for flaws in the transfers. Having worked up a sensation in this direction, he commenced another kind of business. He said he wanted to hire men to work on a farm that he owned in California, and succeeded in hiring two of the Colburn boys who were ready to sell all and follow him. In fact, they allowed themselves to be led captive to a degree that surprised even themselves, when too late. And poor Florence, at first shy and incredulous, was thoroughly carried away by the flattery of her "uncle" who hugged and kissed the unsuspecting damsel in a truly romantic style. But, to make short work of the sequel to the romance, the "rich uncle" left very suddenly and never returned, although the rascal had borrowed money and clothing in the amount of nearly $100 before passing on to other fields.

There was a real curiosity in Dr. John A. Ross's office in Preston. It was a calf or, rather, a pair of calves that nature had made awful work with. They had eight perfect legs and two heads, but only one tail. It appeared as if two bodies had interlocked and grown into and through each other. The heads were at opposite ends of the main body, and the single tail grew out of the side of the center portion. Each set of four hooves pointed in keeping with the direction of their respective head. It lived for half an hour after its birth at Vickerman's farm, and Dr. Ross purchased it after its death. His intention was to stuff it and donate it to the Bennett Medical College of Chicago, where Dr. Ross had graduated. We also heard that the great showman Barnum wanted to purchase it for his show and sent an agent to see it.[*]

A shower of rain brought a bluebird on March 17th—I heard him in the morning. Sap promised to give us a good run, but turned out not to be a very paying business after all. Spring, so long looked for came at last in April, although it was a cold one. Farmers were busy again, the birds whistled and everything looked happy. The weather kept cold and backward yet seemed good for seeding, until a foot of snow fell in late April.

---

[*]   *Preston Republican,* February 6, 1874.

Father had another streak of bad luck — that wife of his left him and took everything she could carry or lead with her to her daughter's house. He came home one day and found her clothes, dishes, and cattle were gone. There had been many disagreements over the years, and David was not an easy man to live with, but we thought this was unreasonable. This began a two-month feud between our families, which included a lawsuit for the property taken. The court found in favor of the wife, and David did not recover the livestock he had fed all these years or the things he had bought for the house. Even after all this, David and Eunice reconciled, and this same bosom companion returned to his house, but empty-handed.

In September of that year Jacob Eyerley, the brute married to Eunice's daughter tried his hand at pounding up father David — he lived in Preston afterward, on public expense in the jail. The following was a poet's way of taking vengeance on the murderous coward.

The Murderous Coward

O shame upon thy coward heart,
O shame upon thy youthful prime;
When none were near to take an old man's part
Thou wouldst commit this horrid crime.
Not guilty! When did sneak do more
Than at the judgement bar to plead,
That justice might this once give o'er
And mercy hide the bloody deed?

In slumbering dreams dost thou not see
An old man, weak and gray,
Struck down before thy murderous arm
In terror and dismay?

Thy murderous fingers grasp his throat,
Thy murderous eyes do glare,
Thy murderous oaths and murderous blows
Sound on the evening air.

O pity for the human race,
That such fell fiends were ever born
More pity for the hapless place
Where brutes like thee are let to run.

Father's domestic warfare gave him no end of trouble and I was induced to take a hand out of a sense of duty. And here let me deduce a theory on interfering in other people's affairs: there is a fine point, an imaginary line, as it were, between sympathy and interference. I love the former, the latter I hate. I pledged never again to meddle with another's affairs knowingly. But, here I was lured beyond the line to help Father out of a trouble that he got himself into against my earnest protest, and I have never ceased to regret it. My father was a kind neighbor and honest to a fault in his dealings, but his family life was made unbearable by frequent "brain storms." He married an old foolish thing that was dominated by as tough a crowd as civilization ever tolerated. This, and the aforesaid brain storms, caused the trouble that culminated in his being set upon and nearly killed by a rough, human brute. The whole affair brought out from my pen "Spots and Shadows," a prose production of no merit. The facts were that David had blundered headlong into matrimony several years back and found himself in bad company, which so disgusted me that I chose this style to express my failings; 'twas a shameful event.

— o —

Spots and Shadows Which Have Crossed

This is the "prose production of no merit" that was inspired by David Cumming's matrimonial affairs.

Now it came to pass that Betsy, David's wife, conspired against him and put him away in his absence while he sojourned in a far country. And when David knew it, he was troubled; and his countenance changed, for he mourned greatly.

And when Rosie, David's son's wife, saw that his heart was sad, she said unto him, "Why dost thou mourn for the land of thy nativity—is not the home of thy kinsfolk fair as the Garden of Eden? Choose ye, therefore, a portion among us and dwell in the land."

And David answered and said, "Thou hast spoken well, my daughter, but my soul mourneth for Betsy, for behold, she hast conspired against me and put me away. Therefore is my spirit wholly gone from me."

Then answered Rosie, "Why shouldst thou sorrow for Betsy, seeing she is not worthy of thee? She hath procured a writing of divorcement, therefore thou art free. What doth hinder thee from choosing a wife from among the daughters of our people? Are they not of a fair countenance?"

And the saying pleaseth David. Therefore he said, "The field of Nathan the Lawgiver, which is on the creek called Frosty, eastward from the park of Cyrus the Hibernian, may be bought for two hundred pieces of silver. I will purchase it, therefore, and take a wife from among these people and dwell thereon."

Now, there dwelt in that country a certain widow called Eunice. And Eunice was a woman of a sad countenance, which was imputed unto her for righteous reasons. But, Martha, her daughter, was a woman of evil repute and a busybody, for she

73

caused trouble between man and man by her much tale-bearing. Now, when Martha heard how David desired a wife, she came and spoke with Rosie saying, "Behold, my mother desireth a home and Joel, my husband, is an austere man and a man of Belial. Therefore, she will not live with me because of my Lord Joel. I am resolved that I will speak to her that she marry." And these things were told David, so David met Martha by the creek which is called Frosty and sought to communicate with her concerning Eunice, her mother.

And Martha wept sore and as she wept she said, "My mother is a virtuous woman and worketh willingly with her hands and, behold, she hath goods and cattle. Therefore, if it pleaseth thee I will speak to her of this matter for I watcheth her mind in regard to the subject."

And David inquired, saying, "How old is thy mother?" Now, David was two score ten and six years old. And Martha said the days of Eunice's years were three score years and two. Then David told Martha how "that Betsy" had conspired against him and robbed him and put him away, and asked her to commune with Eunice for him.

But when the sons of David knew that their father had convened with the wife of Joel the Hydrophobite they were displeased; for Joel was a furious man and did foam, and his spittle ran down upon his beard, and he had antipathy to water insomuch that he drank strong drink and would not so much as wash his body in water. And as often as he was wroth he spoke roughly saying, "Thou zunofferbitch, come nigh unto me and I will darken thy vision and spoil thy countenance forever." But no man cared to contend with Joel, for he was a filthy man and rottenness was upon him.

And the sons of David took counsel together concerning Eunice and Martha. Then said Moses the Younger, "What have we to do with the house of this unwholesome Hydrophobite or the mother of the wife of Joel? For it seemeth to me that David hath grown foolish in his old age."

Quake, who was also know as Phred, answered and said, "Let us have peace, for thou knowest that David is a combative man, therefore, if we presume to dictate, his soul will 'sour' on us. Moreover, as for this woman, she is a stranger to us. Let us not therefore judge rashly, ye cannot discern a tree by its fruit, for an east wind may blast it and an untimely frost may destroy the hope thereof."

Then answered John the Eldest, saying, "As for David, I verily believe his last state will be worse than his former inasmuch as Eunice is more devilish than Betsy, but what will our opposition avail? For David's heart is prone to matrimony, and as for his will, it is according to the laws of the Medes and Persians." So the sons of David arose and went every man to his own tent and spake not unto David, either good or bad, concerning Eunice. Let every tub stand upon its own bottom.

So Martha brought her mother and introduced her to David, and when he looked upon her countenance it pleased him, for David was a melancholy man and he loved to look upon a sad face. Therefore, he covenanted with Eunice and in the seventh month and the fourth day of the month he took her as his wife.

Now it came to pass after this that the house of Joel was in need and Martha said unto Joel, "Behold, we lack bread, go ye therefore into the fields of thy neighbors and labor lest we die of hunger." But Joel answered, "Whyfore should I labor? Behold the field of David groaning under its burden. Go in, therefore, and live, for it is thy portion."

It was the custom of David to walk in his field in the cool of the day and as he did walk he beheld footprints, and here and there the vines were rent and ears of corn broken off. And David looked again and lo, the footprints were fashioned after the feet of the house of Joel. Now, David was a vehement man, and when his indignation was aroused his communication was loud and boisterous like the rage of a fierce and angry storm.

And David went and told Eunice saying, "Why doth the house of Joel occasion against me to rob my vineyard? For, behold, they have broken my hedge and lo, their footprints are found among my standing corn!" But Eunice answered saying, "How knowest thou it is the house of Joel? Per adventure, thy children's children have left footprints as they go to and fro and wander up and down the banks of the stream, for they are fishers." Then answered David, "Can the beagle discern between the bear and the baboon, and can I not distinguish between my children's children and the daughters of thy people? Let thy people beware, for we have law in the land, and it will protect the fruits of my labor."

So Eunice told Martha all that David had said and cautioned her saying, "Beware, lest ye waken the wrath of my Lord David." And Martha rehearsed the words of her mother in the ears of Joel. And Joel was wroth, and the form of his visage was changed, and he did foam exceedingly and he said David is a zunofferbitch, and Martha said dammim.

However, David saw no more footprints and peace reigned between the house of Joel and the house of David for *two full years*.

When two full years had passed, David was taken with a fever insomuch that his life was despaired of—and his two sons, Benjamin and Phred, who is called Quake, did nurse him while Eunice, Martha, and Joel the Hydrophobite watched by his bed. But David feared greatly saying, "They seek to slay me," and he spake roughly unto his wife and commanded Martha to depart saying, "Ye do convert my sick room into a devil's workshop." Then he said unto Eunice, "Thou art aged swine." And Martha departed in fierce anger. And Eunice also was sore displeased. Therefore, she made complaint unto the sons of David saying, "The language of thy father is intolerable." But Quake answered saying, "David is lunatic. If thou canst not endure to care for him, retire to the house of Joel and rest, for we know that thou hast borne a heavy burden." But she refused to go.

Then went Quake and spake unto David saying, "How long wilt thou cause thy children to blush with thy foolishness? Behold, thou hast put us to shame in that thou hast abused Eunice, whom thou wouldst marry against our will. And yet we have treated her with due respect. But thou hast cursed her for doing well." And David confessed, saying, "Eunice hath been faithful." Therefore, Eunice made peace with him. But Martha conspired against David from that day.

And Martha sought occasion wherewith to fill the mind of her mother with evil against David, for Eunice hated David in her heart, secretly. Therefore, Martha communed often with Eunice concerning her lord saying, "David is a miser, he doth feed thee on the coarsest food which thou well knowest. Look at thy raiment also, dost thou not buy it with butter, the fruit of thine own kine? And thy stock is this day starving under his treatment."

And Joel said, "*He that would rob the aged would also steal acorns from a sightless swine.*"

Now, Eunice owned cattle, and David fed them on hay, the fruit of his own field. And Eunice worshipped her cow even as Israel worshipped the calf which Aaron set up, for she was wont to say, "This is my god, oh David, which hath saved me from the mire of poverty and hath caused me to wash my steps in butter." Therefore, when it was told Eunice, behold thy cow doth starve (although it was not so) she became indignant and her heart did rebel against David. So Eunice arose and quit the house of Joel and got herself home to feed her cow.

And while Eunice was yet afar off, lo, her cow saw her and did call after her, in the manner of cattle. Therefore, Eunice being filled with wrath against David because of the words of Martha, imagined evil in her heart. And she came and spake unto David saying, "Why doth my cow suffer neglect at thy hand?" But David answered and said, "Who cutteth up this pork? Thy cow is even now bursting with fullness. Attend to thy duties, therefore, and thou shalt dwell in peace, otherwise it may be ill with thee." Then answered Eunice, "I may not tarry thus with thee." And she went out and departed to the stacks of hay.

When David saw this his spirit arose within him. He took an agricultural implement in his hand and went out and stood at the gate. Now, the weapon which David used was three-pronged and the prongs were steel, tempered by the smith and polished smooth as glass. And David stood at the gate with the weapon in his hand and called unto Eunice saying "Desist, forward woman, for thou mayest not come in hither." But Eunice gathered a lock of hay in her arms and came toward him although the glittering steel stood before her eyes. Neither would she stop until she felt the point thereof pierce her breast. Then Eunice cast the hay at her feet and retired to a place of safety, filled with great indignation. And David lifted the hay upon the prongs of his weapon and bore it again to the place from whence it was taken.

Now Martha, when she heard that her mother had been put to the worst before David, rent her garments in wrath and fury and she went to the house of Phred and said things concerning David. And Phred was wroth at David his father because of the words of Martha. And there was great indignation against David throughout that region, for the people of that land disdain a man to contend with a woman.

And it came to pass after these things that the war between David and the house of Joel waked hot, and Martha conspired with Eunice to quit the bed and board of David. But Eunice held her peace for the space of *one full year*.

Then when a year had passed, the heart of David was sore tried, for Eunice joined with the house of Joel and vexed him sore. And Eunice took the cattle with all that she had and went her way, secretly, while David was absent.

Then the sons of David made themselves strong against the house of Joel. For they said, "This is no other than the work of Joel to gain possession of the cattle that David hath labored to multiply; go, therefore, to let Joel pay for his folly at the Court of Justice." But the judges of the land decided the matter against David. And Eunice and Joel returned home in triumph, but David was exceedingly sorrowful and sore vexed, for Eunice testified in the presence of all the people of the land saying, "Behold, my feet did lack shoes and as for my hinder parts, they had no covering." So David was put to the worst before the house of Joel, and Joel took charge of the spoils.

But it came to pass after a time that Eunice repented and returned in sorrow to her Lord David, and he forgave her saying, "But thou shalt also return that which thou hast violently taken away." So they covenanted, and Eunice returned to his bosom. But the house of Joel was a sore affliction and Martha with Joel, her husband, conspired together to retain the stock which they had taken. Then Phred answered and said unto Joel, "Where now are the acorns and where is the sightless swine?"

— o —

Temperance

There was a great temperance excitement in our country just then, it was called the Women's Crusade and was carried on by women in the form of prayer meetings in front of saloons, the proprietors of which were exhorted to quit their hellish traffic. The thing was continued day after day with indefatigable tenacity until the whiskey dealer surrendered. In this fashion, the war on whiskey was prosecuted with vigor and the number of dram shops grew fewer every day. This furnished us with excitement for about six weeks, when it died down like a great fire and the society known as the W.C.T.U. (Women's Christian Temperance Union) sprang from its ashes. The whiskey war pretty much played out after a few months with not very flattering results for the crusaders or credit to the people, for while the women were treated with due courtesy in some localities, they were followed by howling, vulgar mobs in others.

The papers gave a horrible account of a large reservoir that burst on Mill River in Massachusetts, resulting in untold misery, death and destruction — 150 lives and one million dollars' worth of property lost. Louisiana was also overrun by a flood caused by a break in the levees along the banks of the Mississippi River, and great distress prevailed.

The world was also cursed more or less with war and bloodshed during the year. Spain was still at war within her own borders and in Cuba, and accounts of the killed and wounded reached us now and then. America had little to brag of; the condition of certain localities was most deplorable. I refer to the United States troops that were called to quell riots in the states of Arkansas, Louisiana, Tennessee, and Mississippi. These disturbances were caused by "white league" politicians to intimidate the black race whom they formerly owned as slaves. The state of the country, taken all in all, was rather mixed. Times were hard, money close, and people politically dissatisfied. I hoped a new party would soon take the field.

People were waking up to the cause of temperance all over the Union. A Temperance Society was to be found in every village now and, sad to say, so were many dram shops. When a Good Templar's Lodge was organized in the Waukokee schoolhouse with twenty-five members to begin with, Rose and I were among its members as were most all the young people of the vicinity.

— o —

Frail Bark of Hope

A heavy frost lay on the ground September 15, 1874. It was the first of that fall. The ground was nice to plow but it was still dry under about eight inches.

Suddenly one day, our little baby Reuben sickened. I began to feel concerned for him on a Friday and went for the doctor the next morning. On Sunday, I sent in despair for another doctor. He died Monday, September 21st, about ten o'clock in the morning. This beautiful child passed away before we were prepared to receive the shock. Words are all inadequate to express our sorrow, and if I say he was our best and prettiest child the world will smile and think something like that has been said before. But the memory of that sweet little sufferer will stay with me as long as life lasts.

Requiem for Reuben

A cloud has passed before our eyes,
Our Little Reuben's gone to rest,
For death hath broken tender ties
And torn my darling from my breast.

His mother weeps with broken heart
Above his little coffin bed,
Sure when the joints are torn apart
It cannot cause a pang more dread.

The children silently creep in
And gaze upon his marble face,
They know not that their brother's gone—
This clay was but his dwelling place.

Then little Lewis brings his cup
Filled with cool water to the brim,
He tries to make the babe wake up
And crow and laugh and play with him.

'Tis vain! The little lids are closed
Upon those glazed and sightless eyes,
That burning fever-thirst is quenched
By angel hands beyond the skies.

Little Reuben's doctor bills and funeral expenses amounted to
$17, and it would be ten years before I could afford to buy a tomb-
stone for this sweet baby's grave.

I finished plowing on the bottom in early October, and my next
move was digging taters. A thresher stopped in our yard and
pulled out of our field after a week leaving us 317 bushels of wheat
and 186 of oats. I got twenty bushels of potatoes that year. We had
fine, warm weather for that time of year. I gave Moses Cummings
his notes to me for a horse and wagon valued at $175 and took a
new note from him for $98, payable in two years with interest at
seven per cent. The winter began very mildly. I was sick at heart—
wheat was only 69¢ per bushel—that's what was the matter.
    I fooled away my horse for an old mare aged seven years, three
months. I regretted it almost immediately, and wished I had used
better judgement. Shortly after I bought her, I spent most of a cold
week chasing the mare, who had left my bed and board, and
finally found her after several days' searching. She was also a
"kicker," and one had to be careful and alert passing behind her.
Eventually, though, she proved her worth as she brought me two
good colts, then I traded her for a note that called for $48.37.

We had more blizzards than I could count in early 1875, and
there were about ten feet of snow in some places. We had heard
nothing from our folks in the west, and were a bit worried as the
weather continued extremely cold, about thirty-six degrees below
zero. From January 1st down to the last of February we had every
opportunity of trying the stern realities of a Minnesota winter with

its varieties and variations. I had been trying to do something that winter in the way of cutting wood and getting stones out of the ground for a cellar, but made very unsatisfactory progress. I only hauled a little over four cords of wood to town.

Winter at last surrendered to the superior strength of the ruler of summer and the liberated prisoners of the late triumphant victor now sent up a song of joy and thanksgiving and sprang into new life. The bluebird sounded his first note of welcome March 20th.

Swearing under any circumstances is, I claim, an unpardonable error. But the events of those days were enough to make any man hunt the English vocabulary in vain for a proper adjective to relieve his mind with. In the last days of March, Andrew VanSickle mounted a fast and faithful horse and fled over the border into Iowa like Tam O'Shanter pursued by the minions of the law. He had oats for brains. My guess was that he would stay away. It is said that the way of the transgressor is hard. Everything we saw seemed to verify the proverb. I went to the hovel where he had left my sister and tried to persuade her to come home with me and quit the fellow for good, but my efforts were of no avail except to gain for myself the reputation of a meddler.

Brother Frank Cummings was tending baby again, he had a boy, born April 22nd. The telegraph brought the news that neighbor George C. Passmore had been called to higher life.

Friend after friend departs —
Who hath not lost a friend?
There is no union here of hearts
That finds not here an end.

I was unusually late with my work that season, delayed by bad weather and distracted by my sister Kate's family turmoil. I cleaned up my wheat and found that I realized but eighty-eight bushels of wheat for last year's entire crop, and my corn was a dismal failure. June brought cold, heavy clouds hanging above the northwest horizon, which were driven over us occasionally across the sky, sending chilly scuds of rain across the fields. I completed my cellar wall, quarrying out the rock during the winter. This nice cellar under my house, sixteen by sixteen feet, cost me $30 besides my own labor and anxiety.

New England relatives sent the sad tidings that my eldest sister, Rosanna, died in Boston in July, of consumption. I wish my pen could do justice to this estimable woman who had the care and training of my early years thrust upon her by death's decree — and then a mere child herself, more in need of mother love and care than at any other time of life. She was faithful and true under most trying and provoking circumstances.

> Now many a sweet, sad memory
> Comes surging like a heavy wave across my soul.
> Oh that mine eyes could pierce Eternity,
> This weary, doubting spirit to console
> And from my mind this burden roll.
> Almost I know, yet there's a dread uncertainty,
> An aching void no reason can control.

The end of August brought us more hail and a wet, gloomy time for harvest. We finished binding hay, fishing some of it out of the water. We had as much rain as we needed, and more, I think. The land was so soft that we couldn't drive a horse on it. I feared the grain was much damaged. Very little stacking was done, and the weather continued lowery.

When the weather cleared up, stacking went on rapidly. Wheat was much better than we desponding mortals dared to hope for. My work was hard and discouraging from start to finish owing to weather conditions. My harvesting was done with a self-rake reaper which I ran with one team, and I bound and shocked the grain alone. By September 21st we had hauled our last load of grain to the stacks. Reports of acres and acres of wheat yet unbound and rotting on the ground reached us from the surrounding countryside. It was a pity, as never did better wheat grow, and we were thankful for what we got. Uncle Fowler allowed me 130 bushels of wheat, machine measure, worth $110. I paid $5.25 for threshing my part of the wheat. I put forty heads of cabbage in my cellar worth about $3. The northwest wind made doleful music in the naked forest, and it was quite cold for the season. I husked my corn, getting less than thirty bushels.

We heard that Frank's little Sadie had the sad misfortune to break her arm falling from a wagon. Another sister—almost an entire stranger—died in Vermont of consumption. I received a letter saying that Caroline passed out of earth life as quietly as a babe goes to sleep.

> She bid farewell to earth
> As one lays worn-out garments by.
> Angels record another birth,
> Why, mortals, do ye fear to die?
> Gladly we lay away the old—
> The tattered, faded clothes aside—
> So we resign earth's crumbling mold,
> Its transient joys with pains divide
> For fadeless bloom beyond the tide.

She was a twin and mate of sister Kate. They were separated at our mother's death, being then about nine months old, after which I saw little of Caddie, as she was called. No wonder then that I should write several wistful poems of long ago memories and dreams of death.

Most of my poems commemorate some actual occurrence: a drunken man with blood streaming from wounds in his face caused by falling against a curb stone; being snubbed on the streets of Preston by a lady whom I had played with and worked with in the hay field when we were young. These incidents were forever at work in my brain until the circumstance was jingled into verse, and that is the excuse I make for the "disagreeable stuff" that my hand hath made.

The horses all got the epizootic again in November. Winter had not yet closed in upon us, but it was freezing weather. Warren attended school that winter. I carried him to and from the school-house on horseback or in the wagon on many days. We got our first real blizzard the middle of December. We had a very pleasant Christmas, but the wind made merry with that little dash of snow that fell.

The old year was about to leave and called to settle with us all before going, for it was certain he would never return. My personal affairs were like the restless ocean. Now calm and peaceful, then the waves of affliction or disappointment drove my fail bark of hope upon the breakers of despair.

My health was good the past two years. I wish it were so for my sons. I cannot say any more here about our lost boy. Even so, success to some extent followed my labors. Domestic affairs stood about as usual with me. I was about $20 in debt at the close of 1875. I had "that mare," two cows, seven sheep, a calf, a pig, and twenty hens for livestock. I valued them at $190. My house was worth perhaps $20 more than before, $220 altogether. This winds up my account for the year.

Old year, thy locks are white,
Thy mantle, too, is like the driven snow;
The howling tempest tolls thy dirge tonight,
Thy feeble pulse beats slow.

I look back o'er the past—
A weary traveler, sick and sore.
How many ardent hopes, mere wrecks, are cast
Like rifted ships upon the shore?

— o —

Lines to a Lady Acquaintance

Fair lady, doff thy hat and plume,
And come to earth, once more resume
The plain habiliments of sense,
And deign to hear my meek defense.

You shun me as one would a plague,
But for your friendship I don't beg.
I may be poor — that I allow —
My follies are not small or few,
Still, while in grandeur you may shine
Your heart may be as foul as mine.

In trailing through this mundane sphere
I'm often scant for worldly gear.
My hat gets tattered, torn and dusty,
My coat and pants look old and rusty,
These boots (my best) look worse for wear
Yet I have naught to buy a pair;
But to despise one for his dress
Is merely proud short-sightedness.

Have you forgotten that good rule
We learned while young, attending school?
Honor all men, trust not to sight,
All is not gold that glistens bright;
This rule holds good the wide world o'er,
Fine clothing often hides a sore,
For foolish man, so prone to evil,
Oft wears a shoe to cheat the devil.

But Nick is bound to have his dues
In spite of feathers, hats or shoes,
His brand is deep burnt on the soul
So pray, dear girl, don't play the fool.

— 1874

News of 1875

The papers brought accounts of two very sad accidents or calamities: viz., a powder manufacturing plant near Cleveland blew up, and the states of Pennsylvania and New Jersey suffered heavily from spring freshets. Both calamities probably destroyed $2 million worth of property. There were accounts in the *Toledo Blade* of destructive tornadoes in the south and west parts of this nation, and a fearful cyclone in China in which thousands of lives and millions of dollars were lost. Great excitement was aroused by extravagant stories of gold in the Black Hills, a region located in the southwestern part of Dakota Territory.

I sent $2.25 to the *Blade*, which ensured its visits to my family for another year along with a book titled *The Life of George Washington*. Pennsylvania had a big fire in the wood districts, attended with great loss of property. The city of Osceola was burned with other smaller towns, and the loss was estimated at $2 million. While the East was thus suffering from the flames, the West was being devastated by the pest of the last year. Missouri was being overrun and eaten up by locusts, and other sections were doomed to suffer more or less from the same cause.

Mrs. Lincoln, President Lincoln's widow, was taken to the insane hospital. When J. W. Booth sent his bullet into President Lincoln's brain he laid the foundation of insanity in the brain of his widow. She stayed under treatment for this trouble over the summer, and was discharged by year's end.

I suppose that I should note here an event which caused great excitement in the land for six months, the trial of H. W. Beecher for adultery. This truly great man was subjected to one of the most searching investigations ever instituted. But, after a six-month trial, the jury failed to agree, nine of the twelve favoring his acquittal.

More Klu Klux outrages were reported from the state of Mississippi. Texas was visited by a tornado of the worst kind. The city of Indianolia was entirely destroyed — 300 houses were blown down, 400 lives were lost. Galveston suffered a loss of over $4 million and other towns along the coast suffered in proportion. It also appeared in the news that the natives of the Fiji Islands were dying of a plague that had raged there for four months.

President Grant's seventh message came before the people full of wisdom and patriotism. These points were especially recommended to the earnest attention of Congress:

1st. The states should be required to give every child within its borders opportunity to acquire a common school education.

2nd. No public school should be controlled by any religious sects.

3rd. Church and state should be forever separate and church property should be taxed as other property.

4th. Drive out licensed immorality such as polygamy and importation of women for illegitimate purposes.

Our national affairs were very quiet just then. There seemed to be a disposition to live and let live throughout the country. Preparations were rapidly going on for a grand national celebration the next year, our nation's centennial, and as the next year was also to be our presidential campaign year, we looked for stirring times. Mr. Grant proved to be such an excellent ruler, that many people imprudently urged he be renominated for the second time. But he very sensibly refused to be a candidate again.

— o —

Family Portrait

In 1876 the weather opened mild for the most part with now and then a slight snow. Sleighing was very scarce. "What a warm winter we are having" — this is what we heard from everyone we met. "We don't have snow enough" — this is all we complained of; strange, that men are never quite satisfied!

February's weather brought ice. No great wonder then that Mr. Kistle broke a leg slipping in his own dooryard. We had four days of good sleighing, then we had a freshet and another big flood. Conkey Brothers lost their dam in the scrape. The wind had a "big fun" in honor of St. Patrick and snow lay thick and heavy upon that sheet of ice we had. Our March storm reached as far as Memphis, Tennessee — they got ten inches of snow.

Spring began to make her power felt in April, although she had a job to loosen up that ice. Brother Frank left in search of a better place further west, and returned a month later, having taken up a homestead claim in Nobles County, Minnesota.

A fearful tragedy was enacted not far from here, a woman became insane and killed her husband. She hit him about the head with an axe because he wouldn't get out of bed one morning. Their thirteen-year-old son stopped her and sought help for his father, who lived for about seven hours after the attack. The woman acknowledged that she killed her husband, but did not seem to realize the enormity of her crime. She had been subject to attacks of insanity for more than twenty years, and those who knew her did not consider her accountable for her acts. After a medical examination, she was taken to the asylum at St. Peter.[*]

We had another exciting incident in Minnesota. A band of eight robbers attacked Northfield, and after killing and being

---

[*]   *Preston Republican,* April 6, 1876.

91

killed beat a hasty retreat, followed by all the town except five very
non-combatant women. Two of their number were left dead on the
street and one more was carried away badly hurt. The robbers
killed one man, but got no money. The man killed was a cashier in
the bank who refused to open the safe. We heard those robbers
fared hard, one report had them all killed or captured.[*] Two weeks
later, we heard two of those rascals had escaped. Then that three
were killed, and three badly hurt and captured. The James and
Younger boys finished up their career of terrorism and robbery
with that unsuccessful attempt, which broke up the gang and
landed three of the Youngers in Stillwater prison for life.

We had our first experience with "enlarged pictures" in
October, paying $5 for one of our "Angel Reuben." I was disap-
pointed at first but, gradually, memory and art came to an agreement and the picture has been a comfort and thing of beauty ever
since. I don't reckon its value in money. It appears to me that there
is something in life with which money cannot be compared. This
portrait, a life-like figure, hangs next to an old picture of our family
on the wall of our humble home to this day. Rose arranged these
with other pictures, including her mother and two of her sisters,
Lavina and Mary.

Our Family Picture Wall

At the top to the left sits the hypoxia man;
Blunt, surly, concerted and hair-brained is he,
Whose thoughts, like his whiskers, are scattered and thin
And constantly shift like the waves of the sea.

By his side is the partner and hope of his life,
In her matronly beauty and innocent pride;
Oh, man, thou art blest in thy choice of a wife,
And well may be proud of thy place by her side.

Master Warren, an honest, intelligent lad,
At the left of his mother sits sober and still;

---

[*]   *Preston Republican,* September 14, 1876.

While Lewis beneath him, mischievous and mad,
Exhibits a restless, indomitable will.

Still downward sits Grandma whose silver-tinged hair
Claims that tender regard due to those who have given
Their comfort, strength, talent and patience to rear
The youth of our land in the status of Heaven.

At her right is Lavina, a specimen rare
Of health, strength and beauty — oh happy estate —
May it ne'er be displaced by the tyrant despair;
Long, long may you flourish "Lavina the Great."

By her side, in the corner, the servant of all,
Sweet Mary, the spirit of kindness we see,
With a heart ever ready to answer the call
Of the sick and afflicted, wherever they be.

At the left of the picture, and opposite Lew,
His mother brings Warren again on the scene;
'Tis a fond mother's art forms three babes in a row:
Young Warren and Lewis with Reuben between.

Oh Reuben, sweet cherub, we never again
Shall see that dear babe we so loved and caressed;
But 'tis comfort to know thou art rescued from sin
And to feel that thy soul is forever at rest.

Fond nature, in blindness, would sorrow for thee
And sigh to caress that sweet gem of yore;
But reason well knows 'tis far better for me
That my fair, sinless child should have passed on before.

Now the white robe of winter is spread o'er thy grave
To shield, as it were, thy frail form from the blast.
Oh, say, dost thou ever look back o'er the wave
And witness the gloom that death's shadow hath cast?

We had quite favorable weather through October, but the fates did not smile on my poor soul that summer. I raised fifty bushels of wheat more or less, and thirty bushels of potatoes that were already rotting. Corn and beans came in less than other years too.

A Grange was organized in Waukokee consisting of twenty-six members. We all joined, of course, and it ran well for a season, then the inevitable fizzle as members moved away. We sent our first order for merchandise to Montgomery Wards & Co., of Chicago, through the Grange. These goods came at last, giving good satisfaction. We also had a most thriving Lodge of Good Templars here of some eighty members.

Sister Kate sent up from Iowa for help to come home. This time Frank went to her aid, and again she disappointed us by following VanSickle to Wisconsin after a few months. She returned here the following October. We hoped VanSickle would stay wherever he was, and leave sister Kate in peace. Here, again, I have to confess that my sympathetic nature came very near drawing me into a miserable family quarrel. I wrote the poem "Mankind Is Hawk-Kind" as a way to express my disgust over the situation. To explain a little, the hawk character commemorates a pet hawk we raised one year. While he was a great big harmless fellow whose only game seemed to be mice and gophers, I took the liberty of imagining a more sinister and devious aspect of his species' nature.

— o —

Mankind Is Hawk-Kind

Some men may doubt and sneer and mock
And call this fabled, fancy verse,
Good English from a bloody hawk
Nobody ever heard, of course.
But dare you scoff at man's ability
To learn the language even of a tree?
Then doubt not, neither cavil more
That I am versed in buzzard lore.

Of late a fancy calf of mine by chance did stray,
Which greatly worried me and drove my sleep away.
So I was wandering through the wilds at break of day
And passing by a neighbor's yard upon my mission bent,
I saw a hawk with lightning speed swoop down with foul
     intent
Upon a busy, unsuspecting mother hen.
Quick as a flash I took the situation in
Nor paused to think one moment what to do—
A pebble from the brook with deadly aim I threw
When strange fatality (at least a most unusual thing)
I broke the thieving rascal's wing.

The hen escaped, was little hurt,
Frightened, of course, until she scarce could go.
The hawk lay wallowing in the dirt
When with my staff I thought to finish him,
But better feelings crossed my mind
Which stayed the well-deserved blow,
It was to let the wounded culprit go
If he some good excuse could find
For treating a defenseless chicken so.

So, gathering up the wounded bird,
I put him on an neighboring tree.
"Illustrious thief, if now you have a word
Speak out, I'll listen," said I, patiently.

The hawk assumed that injured look
Which desperadoes always wear,
And like a rascal undertook
To make our motives similar.
"We are rogues well met, sir," he began,
"But mind my wing and give me back my prey
Then take this vile reproach away."
"'Tis time to talk of judgement then.
For on my soul, your business here,
Had I not come to interfere,
Was while that farmer sweetly slept
To kill and lug away that very hen."

"But I'm her ever-watchful lover
And sheltered her with timely cover."
"Culprit," said I, "you're worthy of the name you bear
Thieves never need to stop an answer to prepare;
No rascal ever breaks his country's laws
Without a good excuse and double cause.
But there's no doubt that he who owns these grounds
With half an eye could hardly fail
To guess who meant to pick the bones
Of this unfortunate female.
Your talons in her tender flesh were pressed
With full design to tear the heart out of her breast."

The wretch replied, "Your talk is vain,
Must I be tried by courts of men?
Those who by nature claim the right
To kill my inoffensive race at sight?
Rivers of blood your guilty tribes have shed,
Life in your hands hangs on a brittle thread.
My conscience is on pivot hung
A balance sensitive to wrong,
So let folks keep their noses out
Of business they know naught about.
We hens can regulate our wives
As well as any man that lives,

I can protect my own and will
While I've a toenail left or bill.
You're interfering, chafing, blowing,
That hen and I were merely wooing.
I used my claws in fun to try her pulse
So take your meddling fingers somewhere else."
Then he shook his mangled pinion as if challenging reply.
"Rogue thief," I said, "The whole world knows you lie.
You presume to call your passions love,
But if buzzard love it might have been,
I'm very doubtful then, by Jove,
If I would like to be *your* hen."

—1876

Centennial Fourth

They were having lively times down in Washington in 1876. President Grant was interviewed by a committee appointed by a Democratic House of Representatives, who could find nothing against him but absence from Washington during the summer heat. Mr. Grant politely informed his investigators that he was capable of minding his business, as every president from Washington down had done.

The Republican national convention met in June and nominated R. B. Hayes and William A. Wheeler as presidential candidates. The contest was quite close between Bristow, Blane, and Hayes, which resulted in favor of the latter on the seventh trial. The party was jubilant and confident of electing their man. The Democrats nominated S. J. Tilden and T. A. Hendricks for their men.

There was lots of fun in the air that summer over what George Washington did a hundred years before. The Centennial Fourth was a great day with us, except that the sad news of Custer's death came to us while celebrating. General George Custer, brother Frank's brigade commander, fell in battle with the Sioux Indians during the last days of June. General Custer and 300 men rashly charged into a deep ravine filled with 3,000 Indians and were all killed. Not a man was left to tell the fearful tale. General Custer was one of the bravest and most successful men that ever led men to battle. But, alas! He overestimated his power, and met a sad, sad fate.

Oh, what shall we say of this leader bold,
Whose fame outshines the dashing knights of old?
His followers in other days tell o'er
The daring charge upon the southern host;

And down the manly cheek emotion's floodgates pour
Most heartfelt sorrow for the hero lost.
Oh sad the thought that thy good sword should fall
Among these wolves, and sadder still the call
For help in this ill-fated hour.
We turn away, for who could bear to contemplate thy dying
throe?
Imagination paints the horrors of the scene that we shall never
know.

Another set of presidential aspirants took the field—one Peter Cooper and friends, though this was small fry. We knew the main battle would doubtless be between Hayes and Tilden.

Our presidential aspirants were getting their fill of abuse that year, and for the first time during our nation's history we were likely to lose a presidential contest by means of violence and fraud. The presidential muddle was finally decided in favor of Governor Hayes, but Republicans made a lot of noise over it. The other side talked war, and they had fun sparring over the election results. Many public men brought disgrace and shame upon themselves in their zeal to win this election, and many others won the everlasting respect of their countrymen. Among the latter, U. S. Grant and R. B. Hayes stood as bright and shining lights to future generations. Assailed as they were by every weapon that malice, envy, and treason could bring to bear upon honest men and patriots, they steadily maintained statesmanlike judgement and unflinching integrity. This was the last time I ever voted for a Republican candidate for that high office. For then it began to dawn upon me that the Republican party was not a "prohibition crowd" even though the president-elect was strictly temperate.

—o—

Phoebe Myrtle

With the exception of baby Reuben, Fred writes little about his children. He himself makes a note of this much later in his journal. In Fred's case, I don't think it reflects a lack of caring — to the contrary, family members who knew him depicted him as sentimental and sensitive. I suspect that he had opened himself up so much with baby Reuben that the pain of his death made him more cautious with the others. Family lore also suggests that at least one or two other Cummings babies died at birth, and there are several references in the early parts of the journals about burying a "wee babe" — though no clear note of whether it was Fred's or a neighbor's. Infant death was fairly common in those days, and babies frequently were not even named for several weeks. Now, a daughter is born to Fred and Rose, and she will grow up to be my great-grandmother.

January of 1877 brought a visit from my mother-in-law with all the appurtenances thereof. January 8th brought me a daughter, 7¾ pounds. We named my oldest daughter Phoebe Myrtle, but mostly called her Myrtle or Myrtie. My wife's people took their departure amid frost and snow and danger in the air, and we were relieved later to learn they arrived home safely in Dexter.

The robin and bluebird put in an appearance on April Fool Day, and the snow and ice started down the creek immediately. The plow and agricultural implements were brought into requisition again. Some youthful wolves were being brought to light and destroyed vigorously — 114 were taken already.

There was a little excitement in Preston just then, as Sheriff Peterson's boarders were making it lively for him. What happened was that Peterson's boarders "fell" through the bottom of the prison by prying up boards and an iron plate, which enabled them to escape through the cellar. A young Lenora burglar was caught in a haystack, and Jacob Knudson was brought back to jail by his father. Knudson had killed a man at Fountain by striking him in the head with a board during a drunken fight. Though scores of citizens took up the search for the other prisoners, two of them were a total loss, never to be seen again.[*]

Rose's sister Mary made her home with us and hired on for the summer to work in C. A. Wheeler's woolen mill in Preston. Several months later, that wool factory put a mark on poor Mary and Wheeler promised to foot the bill as far as lay in his power. Her hand was so badly mangled in a machine that two fingers had to be amputated.

Rose and Mary went to Mower County where their parents lived. Mr. Grant went to Europe. I looked forward to the time when both Mr. Grant and Rose returned safe and rested but, personally speaking, Mrs. Cummings was of greater importance than the general.

In May, Phoebe Myrtle got her first tooth and I finished seeding. My brother, John Cummings, emigrated to Cottonwood County, and I still entertained thoughts of going too. My health during the summer was rather poor; rheumatism, Dr. Wilson said. Myrtle was also quite sick, and we worried over her, but she got better after a few days. Dr. Ross, too, was very sick, and people were whispering. He died of alcoholism in June, and was buried four days later.

I was trying my luck with bees again, as I traded one sheep for a swarm that a Norse man had in a box. I read of some men that had attained so high a degree of perfection that they could take up bees without being bitten. We took about twenty-five pounds of honey from our bees, and I got two new swarms the following year.

All Fillmore County was invited to meet in a mass meeting presented as the "Greenback Convention."[†] Its instigators called it "The people's protest against a bond-holding aristocracy." It was really time that steps were taken to head off our money rings. They were a set of gamblers who cried out for our country's honor but who wanted to line their pockets first, and that at the expense of honor, country, or any other thing. Political parties were almost

---

[*]   *Preston Republican,* April 19, 1877.
[†]   *Preston Republican,* October 18, 1877 and October 25, 1877.

broken up and it did not surprise me to see men step out on one side or the other according to how they felt upon the financial question. As for me, I desire to have it noted that I abstained. As it turned out, that Greenback rooster died in the shell — Democratic thunder killed him. That is the way that everything turns out that democracy has a hand in.

I managed to keep my debts pretty well up on a line with my income that year. My stock was slowly but steadily increasing, but I found that when goods were increased, they were also increased that eat them.

Lavina had a little feminine blacksmith put in an appearance January 12, 1878, weighing four and one-half pounds. A right smart chance for her to make something of that, I reckon, as there was but very little chance for original sin there. Couldn't possibly have been more than half a gallon, for a pint's a pound the world around.

I undertook a job that winter that my neighbors said would inevitably result in failure, viz., preaching temperance to our Norwegian neighbors. But I didn't know when to give up as long as I got a hearing. I espoused the temperance cause, and began a series of temperance debates at the Willow Creek Schoolhouse No. 131. I urged the boys with all the reason I could command, to save themselves from this devil drink. Some of them I know disregarded the warning and went to the bad. Even so, our missionary work for the temperance cause proved a success, I think. After a few months, we closed up our labors in the Norse district with the best of feeling.

March looked like spring, grass was growing, robins and wild geese were whistling, and the voice of the turtle dove was daily expected. A young cow was born, and we commenced spring work. We got Myrtle's picture taken to add to our portrait wall. The 29th of March was a day long to be remembered by me, as it was then that I discovered my little five-year-old Lewis was blind in his left eye. One doctor advised me to go immediately and have it operated upon, but better counsel from Dr. Jones was "Save the good eye from bungling doctors," which I did.

Aunt Lydia was sick, too, stricken with paralysis early in May. She first complained of blindness or dark spots before her eyes, then came the benumbing stroke so dreaded and so dreadful. She was bed-ridden, which began fourteen years of close confinement. Dr. Jones was also an unfortunate man, as his wife died and he hardly knew what to do.

> Pity is a hollow thing,
> For though we know our neighbor's woe
> Yet little comfort can it bring
> To him whose soul is burdened so.

Sister Mary stepped in from Dexter, and decided to try her fingers in Mr. Wheeler's woolen mill again. A whiskey fight happened in Preston[*] and it would cost $1,000 to the county to try the outlaws, and still the real outlaw was left to run: whiskey was the entity that should be arrested, murderous whiskey.

We returned from our yearly pilgrimage to Dexter to see my mother-in-law. Everything was lovely there, but the country was not as handsome as here. December brought us delightful weather, but hard times. Our wheat brought the dismal price of only 35–50¢, corn was 20¢, and I had about 200 bushels of ears; potatoes, I had none.

It appeared to be a time of trouble, both of things human and things in nature. Blight and famine prevailed all over the world. While war and pestilence did their share toward making old Earth a chaos, men appeared to have overlooked their relationship and common brotherhood, and slayed each other like brutes. Money, position and power were all the excuse men required to spill the blood of fellow men. Money enough was spent and destroyed in transporting armies and conducting military operations to entirely relieve the wants of suffering humanity.

"Hard times" was a common complaint. There was no money anywhere to be had. Some claimed it was owing to surplus of breadstuff, but I supposed the real cause to be the unsettled condition of our money matters. Our greenbacks were scheduled to be redeemed in coin January 1, 1879, and moneyed men naturally felt

---

[*]   *Preston Republican,* June 13, 1878.

anxious about whether the government would be able to cash up and, lest it fail, preferred to keep their stamps within reach. If this was true we reasonably expected business to brighten up with the ushering in of the new year.

— 0 —

Fasting and Prayer

Glory, R. B. Hayes took Mr. Grant's place in the White House in Washington, D. C., in 1877. Mr. Hayes's inaugural address was exciting general interest just then—it was calm, clear and patriotic. Mr. Hayes would find it hard work to carry out his Southern policy. If he succeeded, it would be a crown of glory upon his brow and, if he failed, people would call him a fool.

Governor Pillsbury called for a statewide day of fasting and prayer because of the plague of the locusts out west. For several years past, the western frontier of Minnesota had been devastated by grasshoppers inasmuch that the settlers had to call in state aid to save themselves from starvation. Those who went from here, like my brothers John, Frank, and Moses, came back to work in the harvest. I thought the governor's strategy was folly, but it turned out that our people did not suffer as much from grasshoppers that season as in previous years.

The first great railroad strike with all its attendant horrors broke out in July. These strikes had become a common occurrence and also a common menace to peace, prosperity, and even life itself. Riot and murder was a very mild way of expressing the disorder that prevailed. The cause of all this was the reduction of wages on the railroads. Railroad employees refused to work first, then forced those who would take their places to quit and, finally, burned property and killed everyone who opposed them, even fired on U. S. troops. But, eventually, the riot was squelched.

Nationally, the Greenback party began an agitation against retiring the greenbacks, but the banking interests combined to carry the measure and, by power thus gained, ruled the country with a rod of iron ever since.

All was quiet within our border during early 1878, but Europe was very much excited. The Russian blood was up and other nations were trying to stop her victorious legions from taking the Turkish capital, Constantinople. Russia was not altogether honorable, and yet if any nation ever deserved a sound threshing it was Turkey. England sent six war vessels to Constantinople against the wishes of Turkish officers, and Russian troops began to move into the fortification of that city as much as to say "hands off, Johnny." So Johnny pulled back again. All this was done in apparent friendly feeling, but everybody knew what it meant. It meant, simply, that England didn't like the look of things and Russia didn't care whether England did or not, while poor Turkey didn't like being eaten by either one. So Turkey gave Russia everything that she demanded, which just about annihilated Turkey in Europe and debarred every other power from using the straits for their navy war ships, while it left the said water free to all merchant vessels.

As for ourselves, we were working into hard money. Bravely, some knaves, or fools, in Congress demonetized silver in 1873, which law was repealed by a two-thirds vote and silver money rattled around quite lively. R. B. Hayes was between two fires and yet he remained unmoved. Mr. Hayes had vetoed the silver bill and heard something drop inside of an hour—that was the quickest time on record—and everybody was glad but him. If my advice was good for anything I would say that Mr. Hayes had better run home—he was altogether too honest to live in Washington. But, once things quieted down, we concluded that Mr. Hayes was not so bad a man as he might have been. A Democratic battery opened up on Mr. Hayes, but the rebel guns were like the sportsman's gun that aimed at duck or plover; scared the game and kicked the gunners over.

The Southwest was being visited by a fearful epidemic of yellow fever, which raged worse than ever before on his continent. People fled for their lives and the sick were left to die while those at a distance dared not go to their rescue. The death rate ranged from fifty to seventy-five daily in New Orleans, and about a quarter of those taken by the disease died within forty-eight hours. Minnesota had a sensation as five large flouring mills in Minneapolis went up due to spontaneous combustion, with sixteen lives lost.

William Cullen Bryant, America's poet, died in June of 1878. Aged and gray, he was eighty-four, and left a record that the ages could not blot out. I loved his poems, and in my school days committed many of them to memory.

— o —

Rigid Economy

Fred is thirty-two years old at this point. He is working hard, watching his family grow, and seeing old friends die. He starts writing about his own mortality, and is concerned with doing the most and best he can with the time he has left. This becomes another major theme in his journal and his poems, which is ironic, as Fred lived to be more than ninety years old.

The new year of 1879 came in very squally and boisterous. The mercury went down to twenty-odd degrees below zero. Everybody said we were having splendid weather, with about ten inches of snow. I traded off old kicking Bet and sold another hog. Eight inches more of snow fell in the middle of January and the mercury stayed down between twenty and thirty degrees below zero. Then the weather turned decidedly splendid, our snow went off and left our travelling decidedly horrid with the air full of rain and mud on the ground in early February.

Our neighbors over the water were much alarmed over the presence of a terrible disease in Russia known as the Black Death. This was rapidly spreading over the empire with more fury than our yellow fever of last fall; for, as its name indicates, it was almost sure death and it made a clean sweep as it went.

Rose and I renewed our membership in the Preston Grange in March, though we found it to be in a fizzling condition. I never got much satisfaction out of it for, although its pretensions were laudable, its managers and those who persisted in running it were either incompetent or self-seeking demagogues and just as dishonest as any other set of political bosses. The most conspicuous leader among them, and one who had the most to say against salary-grabbers and dishonest public men, tried to trick me out of pay for shearing a few sheep for him, but I turned the trick on him by "shear luck." Exit Grange, exit leaders.

That hay we had gathered the last summer was intolerably bad for stock, it made the horses slabber and the cattle look sober. We hoped for an early spring so we could turn the livestock out to pasture. The robin said spring came on March 8th, and if any more evidence was required, the mud put in its irrefutable testimony. We had some very warm weather in April, and vegetation was striking out rapidly skyward by mid-May. We had a tough time for bees with the dry and very changeable weather, cold one day and then hot like Milton's hell. But it finally turned to very favorable weather, and wheat and grass took new courage and a fresh start.

Mrs. C. went to see her pap in Dexter. This time I tried to keep house alone, and it was useless to keep track of the young ones she left, so I just called 'em up and fed 'em twice a day and let them range. There was no law restraining such stock, though our town had voted to restrain cattle that year.

Mrs. Colburn and Florence went to Oregon. Mrs. Moore, an old Vermont friend of my father and mother died in April in Vermont. She was one of the few whom my memory recalled fondly away back from early childhood.

They are going, going one by one,
Across the silvery strand.
And I should soon be left alone
But for the fact where they have gone
I, too, must go — the time's at hand.
Nor can I dread the change,
For gathered on that other shore
Are dearer friends than here, and more
Whose thoughts and love and labor have a wider range
'Tis true, and I'm content, although it seems strange;
I'm weak, I've gone astray, I've sinned,
And yet methinks they know, that band,
How I was cast all helpless on the treacherous sand.

Almost a month passed since my last scratch of the pen, and all for lack of time. My cow died, a wolf killed my lamb, and the old turkey hatched thirteen chicks. I attended to Mrs. C. after her return, for she was ill.

My bees were doing well, and there was such an abundance of white clover as I had never seen before. I got three gallons of juice in the honey business. Also, I bought a queen to replace one which I killed by some mishap, but then learned that I played the fool again. That hive had a queen — I came near losing a dollar by too much haste in judging a vacant chair in the royal family. But, by hastily removing the usurper, bloodshed was prevented and an already exhausted treasury saved from bankruptcy.

Scarlet fever kept people in and about Preston in continual anxiety for their children, while further out, yellow fever was again heard in the land in July. Memphis, Tennessee, was the most afflicted place in the Union according to the papers, and no man cared to go there.

Preston people said they were sure of a railroad, as a narrow gauge was promised to be here in one hundred days. The golden glories of harvest were upon us again, and wheat was threshing out between eight to twelve bushels per acre.

Cattle bothered some that fall on an account of a change in our law the previous spring. A fool in the neighborhood pulled down his fences and forgot to pasture his stock. Now he was cursing his hogs, looking over a ruined cornfield, and declaring his neighbor's cattle did it.

By December 21st we had snow enough for good sleighing, so of course, our young people had a merry Christmas.

Greenbackers were trying to rally force enough to sustain a club at Waukokee, but it looked like an abortive attempt. Greenbackism was going to die out, I believed. Nothing but a speedy and effectual dose of "rational" would save the party. Its adherents were made up of those who were in debt, the dissatisfied, and those who liked to live high and work little. This element would kill any party.

That narrow gauge landed in Preston, so we had it at last. It had been spoken of several years back, but was not built till that fall. I helped halve some timbers for its construction, "spiles" they were called, and were bought from Uncle Fowler. I stopped at the narrow gauge railroad depot and declared it a success. Preston had begun a new era, that was my decision.

And now came the end of another year. Oh, how fast they flew away! And how those rolling years reminded and admonished me that I had no time to waste, and how hard I studied to make the most of the span allotted me. As for my own prosperity during the year I had nothing to boast of, nor could I complain. I had, by rigid economy, kept up with the times and paid up half of my last year's debts. I did quite a business at bee-keeping, but suspect that my real knowledge of the trade was very meager although I got the desired results—honey—in considerable quantities. My hay crop was abundant and good, while other crops were but light. My property had not increased very alarmingly, nor had it shrunken very much. One good thing fell upon me on this farm—I always had plenty of work, both winter and summer. This kept me out of mischief which, by the way, was a consideration.

The wants of my family increased somewhat, but as the children grew it was less trouble to care for them, and so I suppose the one overbalanced the other. My children are rarely ever mentioned on paper, but they occupied much of our time and were a source of great concern to us. How to feed their souls as well as the proper care of their bodies was a problem requiring constant study. They appeared to be healthy, mind and body.

I have to confess to writing some very poor poetry in the fall of that year—spite work, mostly, caused by a neighborhood quarrel over unruly cattle and a general misunderstanding. Writing seems to be a way I had of quieting my agitation whether in grief or anger, a sort of safety valve as it were.

Goodbye, old year, you cannot stay
Your mission's finished and why need you linger?
And so, with us we can't expect delay
When yon pale marshall beckons with his finger
And ushers us into eternity.

—o—

Repeaters

Things were boiling hot in Washington D. C., Mr. Hayes did a pretty good job in the shape of a veto. A rebel Congress was trying to choke the life out of our nation by withholding necessary supplies from our army and navy until certain conditions were complied with; the conditions were that said army and navy shall in no case be used to prevent disturbances at the polls. This, of course, meant that our erring brethren proposed to bulldoze Negroes with none to disturb them. Blaine of Maine and others proposed an amendment to this bill stating that no man should dare carry deadly weapons within a mile of any voting place on any election day on pain of fine or imprisonment, which was rejected. This showed that the rebel element predominated. Another notable event was the exodus of Negroes from the South. They were leaving the South by every way and byway, and may the Good Lord push forward the work until not a Corporal's Guard could be mustered south of Mason and Dixon line.

William Lloyd Garrison, the world-renowned Abolitionist was dead. The rebellious sons of Dixie continued to stir up trouble. One candidate for sheriff was shot thirteen times in the back because he dared to accept the nomination of sheriff of Yazoo County, Mississippi.

Politics were so mixed up that I made up my mind to get out of the snarl, take a breathing spell, and then try and find out who it was best to support. After looking the situation over, I was half persuaded to rewrite what I did two years ago in regard to political matters except that I thought I now knew what to do. I expected to vote a split ticket that fall, just Republican enough to take the curse off and Greenback enough to express my disgust for the third term candidates: Pillsbury, his policy, and repeaters. Generally, by repeaters I mean renominated candidates; I wish one term of office made a man ineligible to the same office for the next term. This, it seemed to me, would have a tendency to discourage wire-pullers who were looking for puppets. Pillsbury was running his third heat for the governorship of Minnesota, which I considered an insult to our people.

Mr. Hayes again did a blind, blundering piece of work in the shape of a recommendation to Congress to retire or destroy what few greenbacks had escaped the wanton money kings' furnace and also to stop the coinage of silver. This was the exact import of his message, but when congressmen asked him what he meant, he professed that he didn't mean hardly anything. But one thing was certain, the people would put their own construction on the act; some said it was a premature discharge and pointed to what we may expect of the next administration if the Stalwarts succeeded in electing their men. Therefore, Mr. Hayes unwittingly put the people on their guard and then, seeing his mistake, tried to pull the wool over their eyes again. You spoke one year too soon, Mr. President.

Well, after another election agony was past, the whole North had gone over to the Stalwarts, or Republicans. Perhaps it was all for the best, but the Greenbackers feared and trembled while the Honest Money party rejoiced and were exceedingly glad. As for me, I was like Henry Ward Beecher hanging on the ragged edge of doubt and despair, hoping for a safe voyage, but I was suspicious of the crew and managers of the old craft.

General Grant went to Cuba and Mexico in the winter of 1880. We thought he would probably be nominated as the Republican candidate at the coming convention to be held at Chicago in June 1880. But Mr. Grant was not nominated, he was excused, and another—and I believe as good a man and one more sure of an election than any other—was chosen as the representative of old-time Republicanism. Chosen unexpectedly to himself and everyone else, James A. Garfield was the man.

The presidential aspirants were three that year, viz., Hancock, Democrat; Garfield, Republican; and Weaver, Greenback—all good men, I guess, to let them tell it. They all claimed to know just what we most needed, and none claimed to have aspired to the place—so I suppose none would be elected?

There appeared to be a queer notion among men that a vote cast for a certain cause carried with it a great moral effect, although the act of doing it weakened as good a cause and drew off that much force from the real enemy's front. For instance, the temperance men made a nomination without the shadow of a chance of success, just for a moral lesson to the Republican party. The woman suffragists went over to the Democrats in hopes to whip the other side into submission, I suppose, for surely they could hope for no help from the Democrats of that day who would gladly disenfranchise four million men in the South, or turn their vote into fetters stronger than the bondage of 1860. Then every vote cast against Garfield that fall would count as one for Hancock, and that meant one against temperance, one against universal suffrage, one against law and order, and one for not only hard money but hard times. That's the conclusion I came to, for the real battle would be fought between these two men.

The political sky looked very squally though we knew November would clear it all away, I hoped, with Garfield elected. I was ready to admit, however, that my choice would have been Weaver, were such an event possible. But it was entirely out of the question, although I hoped to see a Greenback congressman elected from our district in place of our old repeater, M. H. Dunnell, who had been elected so often that he looked like a last year's bird's nest.

I took occasion to chastise those who should presume to waste a ballot on Weaver or Dow or anybody else. But when the hour came for action, I fell into line myself and voted for Weaver, of the Greenback party. Strange to say, though my sympathy had always been with Prohibition, my vote was not always counted that way, for which I have no excuse except that campaign orators with their false representations would deceive even the electorate, and my poem "Which Way Shall I Vote," although a miserable literary production, faithfully described the situation. I didn't mean that my mind had been troubled with regret since the election. Mr. Weaver was a noble man and well worthy of any office, but the principle for which the party was contending was the cause of

justice and humanity and of honesty, the old parties to the contrary
notwithstanding.

The Telegraph was busy carrying the news about the election.
The North, said the latest intelligence, was solid for Garfield and
this pleased me, for I hoped there would be something else to
quarrel over after that. It was a shame that our people quarreled
like dogs while sharpers and ring politicians were robbing them of
their money, their liberties, and their hope of equality before the
law. We had our regular excitement over the election again. The
Democrats were charged with forgery that time, forgery and
perjury in the very headquarters of the party.

— o —

Greenback Rally

One of the improvements slated for the farm in 1880 is digging a well to bring water to the house. It is the first time that Fred mentions that all their water is hauled by hand, in buckets, from the spring. Such routine things were not worthy of note — everyone did that type of hard work on a daily basis. The children probably helped haul water, especially on laundry day. Doing laundry was a two-day job, usually spanning Monday and Tuesday in the Cummings household. Everyone took a turn at the crank to wash and wring the clothes. Hanging the clothes to dry and ironing was women's work, which consumed most of the second day. This was a weekly chore, other household chores such as beating the rugs were done less frequently but also on a regular basis.

Just for an experiment I decided to try keeping track of my daily labor in 1880. I had three hours of chores each day caring for our stock, plus other work to keep the farm business going, and I planned to track this other work for an entire year.

Sickness prevailed, and in my house the young people were all sickly, especially Myrtle. I sneezed at least four times daily. James Bowden returned from Nobles County with Ruth, and reported five of his brother's children were dead in thirteen days' time. Diphtheria was the scourge of the winter, as yellow fever was of the summer. We had a visit from my wife's people, the Howes and the Goulds, who stayed about a week.

I was trying to do lots of business that winter. I had two jobs half done, lathing my bed chamber and digging a well, plus a school wood job on hand. The weather was fine toward the end of February, but I didn't get much good out of it. I was about as miserable as my worst enemy could wish. My side was so sore that I couldn't think of doing a thing except chore about. I finally put myself on the sick list, and Dr. Jones came to see me. He said I had neuralgia in my side. A small calf appeared at the barn, and two or three lambs also put in an unwelcome appearance. They had to take care of themselves until my blister quit pulling. After being subjected to three weeks' siege of lameness caused by neuralgia and Dr. Jones's blister, which latter cost me $3.50, I began to be myself again.

The deaf heard and people could see by lighting—two very important inventions were patented, called the "electric light" and the "audiophone." These things, crude curiosities at first, became real necessities that men couldn't live without.

When February 29th came around, I knew it wouldn't appear again for four years. And perhaps—well, I did reflect sometimes—perhaps I would never see it again or record it on these pages. Oh, how strange it seemed: today, a living, moving, thinking being and tomorrow, an inert lump, changed in a moment to merely nothing that the wondering world knew of; for no one knew for sure if we lived on in spite of nature's changes or if it all ended in nothingness.

I swapped off my pet cow for a breaking plow and numerous other trades such as a knife for whip caps. I found three out of six swarms of my bees dead and thought perhaps I might do better to have them out. The weather was quite favorable the following week, but water ran into my cellar, making me glad the bees were out of it. I now had four cows to milk, four calves to feed, and the sheep had some dozen lambs or more. Uncle Fowler tended the sheep mostly.

April brought a week of mud. I brought my well-digging to a close twenty-one feet below the surface April 1st, having struck solid sand and a vein of water. But indications were that we still had to drill a few feet. Our well was so near finished that we thought another load of stone and an hour's work would put the end to it. That was not the case though, and we had to find a way to go deeper. Two weeks later that well was still a nuisance.

Ever since we settled here we have had to carry all our drinking water from the spring about thirty rods away above the house. We dug a well in hopes of finding water without going down too far, but we were disappointed. We went through over twenty feet of gravel, then six feet of loose sand, all of which would take in all the water. Then Uncle Fowler hired a machine to drill down thirty feet more through sand rock, got disgusted, and we quit the job as a failure.

If I wasn't used to this world I would have been discouraged. It was almost May and no seeding could be done to speak of as the ground was a bed of mud, but the sun was bound to shine sometime. We proposed to raise cane that year. I took to speculating again by way of buying blooded stock in the shape of a bull from Dr. R. Thatcher. More rain fell within the first week of June than fell altogether before during the year. My corn looked cut through the mud and would have done better if the water ceased and gave it a rest.

The weather was nice again, but the newspapers said that it behaved very badly in other places. Fearful tornadoes and heavy rainstorms were the peculiar features of bad conduct up to the middle of June. Mrs. C. and the folks all went off celebrating the glorious Fourth in Lime Spring, with me minding the livestock at home. My bees were making me much bother. I had seven hives full by then, but I had been beaten four times by them. I harvested my honey crop and thought it a fair renumeration for all the bother my bees made, getting probably ten gallons as the outside figure, which I reckoned was as good as a $10 bill. Another honey harvest at the end of August gave us another eight gallons of juice.

People were harvesting by July 25th, and so were the chintz bugs. Rivalry and strife. All the excuse farmers had for their early attack was that the bugs began first. Uncle Fowler made me hire my own help for that harvest, and increased my share to half the crop. Our grain was cut the first week of August, but it was not very abundant. I sold my ponies for $140 and a cow. Spencer Peterson took the colts and gave me $23 in cash, the promise of $17 more directly, a cow worth $20, and notes and a mortgage calling for $100 more.

We had a curiosity in town, and also a monstrosity. Miss Eveline Bowden had been very sick with inflammation of the stomach and she vomited up an animal resembling a snake or worm about one and one-half feet long. Mrs. Colburn and Florence returned from Oregon and we expected Frank and his family soon. Warren was trying to learn the art of music on the melodeon.

I had a great deal of work laid out for October. My house needed repairing, plus the boss said he was going to sow rye for pasture, and I had the work of it to do. Peterson was going to keep my cow ten days and buy two gallons of my sorghum (Myrtie called it sour gum, very appropriately), and work for me two days as a recompense.

Quite a stir was created in our quiet valley by a runaway which would take too much time to narrate fully in my journal but which I will remember so long as I live. Nobody was seriously hurt, but several were badly scared. A woman was carried about two miles over the roughest roads in town by a span of young, spirited, and entirely uncontrollable horses. I played a part in chasing and capturing her team. When rescued, she was nearly bereft of her senses, but other than a thorough shaking and severe bruises, she escaped mostly unhurt.

F. W. Colburn left to try his luck in Oregon in October. We learned to our dismay that he died in early December after an illness of thirty-four days. He never was well after he got there. He cast his all into the balance of fickle fortune, and fate turned the scales against him. Poor Monty, his day of disappointment was past, and he saw with a clearer and surer light. He left a wife and an infant son to struggle with poverty.

I went out to a Greenback rally. Professor C. S. Powers spoke at Waukokee on the way things were supposed to be done. Some truth, some nonsense. This learned professor gave notice to the public that he would speak at town upon the state of the country and promised to tell people how to use their elective franchise to the good of the whole country. The people, accordingly, assembled; for we were a law-abiding, patriotic people and Mr. Powers was a most able speaker. But, the lesson that he taught us he would never teach again. Upon that well-remembered night he began by stating that General Grant was a hero, and that whatever disposition the American people might make of him, his modest manners, undaunted energy, and his sterling integrity had emblazoned his name upon the pages of our country's history forever. We nodded our agreement, as he stirred our patriotic feelings. Then he dropped the general as a subject matter.

He next gave us a lesson on finance, commencing with the history of greenbacks and bonds. He told their origin: how they were created in a time of trouble and financial distress to relieve a nation for the time being, and with the intention as well as the promise to redeem the same in specie. "And now, gentlemen," said he, "shall we keep our nation's honor and redeem her sacred pledge, or shall we repudiate? Shall we insult our creditors by offering worthless paper (greenbacks) for these interest-bearing bonds? Will they not, as creditors, sue these notes — for such they really are? Most assuredly, they will, and collect them too, every cent of them. Greenbacks are valuable just in proportion to our disposition to carry out the design and pledge of their origin, namely to redeem the same in gold or silver."

We, like fools, thought it sound logic and voted accordingly, while Powers laughed at the simplicity of the masses and later declared he was joking. But we wouldn't be fooled again. We decided that if Powers was so apt to say so many funny things, then we couldn't know when he was in earnest, and his reputation suffered. I was satisfied that Mr. Powers lied a big lie according to his own definition of falsehood. Half the truth, said he, is the most effectual way of lying.

According to notes of my work, I had a very busy year of it. Although the last month's record was not very brilliant, I really was busy choring and caring for the stock. I engaged in several projects that failed to pay expenses, digging a well, for instance. My crops were a partial failure too, hardly paying expenses, but my honey crop paid me well. I made some foolish cattle trades that year. But although I lost quite a bit financially, my greatest loss as I now recall was a loss of temper, which called forth some bitter, spiteful verses. Among the trophies of the year were a poem titled "The Shrunken Soul," another called "Jonah's Whale," some unwritten lines that come to my mind sometimes, and other verses suggested to me by some incident or another.

So, summing it all up, I had to confess that almost every project failed me that year. Many perhaps, my own fault, some on account of the weather. So endeth the year's calamities. And here I will stop telling my misfortunes, although these are but specks to what happened to me. Dishonest men laid their snares for my feet and, little by little, I learned to distrust everyone. Even my own flesh and blood loved me only so far as I could be used for their benefit. These thoughts served to make me more cautious in my future course with people. All men are liars, was my new motto. I vowed to keep a closed mouth, make no promises, and close no bargains until I was sure I knew what the other side intended doing, no matter who he was.

— o —

A Year's Work: 1880

Just for an experiment I tried keeping account of my daily labor as well as keeping the journal. I had for everyday chores that winter eight head of cattle, twenty-five sheep, and two horses to care for of Uncle Fowler's; plus four head of cattle, ten sheep, two colts, three hogs and twenty fowls of my own. This occupied three hours daily. Then, hauling wood, fixing stables, chasing cattle and all such jobs took up so much time that I vowed to make a note of them for one year.

January:
Spent all or part of 15 days cutting and hauling wood for
  myself, the school, C. P. Fowler, and Dr. Jones.
Spent part of 10 days shelling corn or cleaning and hauling
  wheat.
Spent part of six days getting lumber, then fixing the stable, the
  house, and building a privy.
Spent four days butchering swine for Peterson, Bowden, and
  us.
Chased colts, dogs, and cows.
Fixed a well-elevating bucket.

February:
Spent eight days hauling wood for us and the school.
Spent six days repairing the chamber lathing and fixing the
  east window. Bent the Sabbath nailing on two bunches of
  lath.
Hauled and rolled stone downhill for four days.
Doctored lame colt.
Went to town with grist, got eight bunches of lath.

Got so lame I could hardly cut my oven wood. Called in the
   doctor to see my rib and nursed my blister. Spent two days
   lounging about sick and sore.
Killed mice in a beehive. Health improved slowly.

March:
Went to Fountain with Mrs. Gould, found a clevis.
Wheeled out many barrow loads of manure from stable.
Nailed on some lath for two days.
Went to town caucus.
Saw a bluebird on March 6th. Opened my cellar.
Spent nine days cutting and hauling summer wood for us, Dr.
   Joens, David, and school.
Went to the mill. Put out my bees, half of them dead.
Fixed things about the yard. Fed bees. Cleared away rubbish.
Spent four days digging a well.
Went to town with butter & eggs.

April:
Spent all or part of nine days digging well, and hauling and
   setting stone in the well wall.
Set out a willow fence.
Spent all or part of five days cutting and hauling summer
   wood. Hauled home two bunches of lath.
Spent five days plowing.
Moved S. Peterson.
Trimmed sheep, fixed a broken pitchfork, and cleaned seed
   oats.
Grubbed up two trees.
Rigged the old seeder, and spent six days sowing oats.
Dug up three trees. Finished walling the well and measured off
   eight acres of land for corn.
Went to Dr. Thatcher's with a heifer.
Cleaned and seeded eight bushels wheat, hauled two loads
   rails.
Went to the mill.

May:
Finished seeding, twenty-eight acres in all.
Tapped boots.
Spent five days plowing and dragging fields for C. P. Fowler.
Went to Thatcher's & bought a bull.
Hauled seven loads of fencing.
Fixed the west chamber window.
Planted potatoes and cane. Chored about for two days.
Spent three days hauling manure.
Cleaned out the well, chored about, and fixed the chamber.
Spent five days plowing David's corn ground, then spent all or
    part of five days more planting and harrowing corn.
Wrote to Frank. Helped drill.
Sheared sheep.
Planted sugar cane and more potatoes.
Broke an axeltree. Went to town.
Fixed our baby Reuben's grave. Divided swarm of uneasy bees
    for fear they might fly away. Cleaned the cellar.

June:
Fixed up my wagon.
Spent all or part of five days helping to build a pasture fence
    and dividing sheep.
Went to town for grist.
Spent three days working on the road.
Got the team shod and cleaned out around.
Spent seven days plowing the corn.
Spent part of five days weeding and hoeing potatoes.
Went to town. Hived swarming bees.
Went to mill.

July:
Hoed potatoes, herded cattle, etc.
Sowed buckwheat and plowed cane.
Spent seventeen days mowing and haying on long field,
    meadow, and down by the creek.
Divided bees and tended my neighbor's cattle.
Went to Carnegie's, hived a swarm of ugly bees.

Extracted about 9½ gallons of honey.
Looked after Jim Bowden's bees.
Went to town.
Spent three days trying to make the old harvester work, then
  engaged a harvester and a man to bind on ours.

August:
Spent six days harvesting oats and going to the mill.
Went to a funeral.
Fixed a rack and spent six days stacking wheat and oats.
Spent all or part of eleven days plowing.
Went ten miles for a sick neighbor.
Took plow and mare to shop, fixed the fork and plow.
Went to Carnegie's.
Extracted honey for three days.
Hauled a load of wood to town on two days.

September:
Spent all or part of six days hauling wood to town.
Hauled fencing for two days.
Spent all or part of eleven days plowing.
Spent four days helping Davis, Bowden, and Olsen thresh.
Spent three days trimming, stripping, and cutting cane. Broke
  the wagon, so went to mill and got it fixed.
Dug potatoes, picked a load of corn, and went to town with a
  cow and calf.
Hauled manure for two days.
Tinkered and chored about due to rain.
Packed one bushel of grapes, went to town for lumber.
Removed rubbish preparatory to repairing the house.

October:
Gave chase to a runaway team and helped rescue a woman
  from peril of life and limb.
Worked for eight days repairing the house; doing carpentry
  and plastering our chamber.
Sunday, visited Reuben's grave.

Went to Bowden's for lime and rye, went to town for flour, got
     100 pounds.
Took two days to finish plowing and putting in four bushels of
     rye on the hill.
Spent three days picking corn and digging potatoes.
Brought my cow home.
Went to town.
Scraped dirt, and husked for two days.
Threshing for Norton two days.
Hauled manure, wood, and made stable.

November:
Fixed the shed and stable.
Went to vote.
Threshed 120 bushels wheat and 275 of oats.
Hauled school wood for four days.
Scraped my dooryard and cleaned up a grist.
Chored about; helped Facey husk and thresh for two days.
Banked the stable and house for six days, trying to keep cold
     out.
Went to town.
Spent four days butchering two hogs and a beef. Salted pork,
     took a hog to market, and cut up beef for Uncle Fowler and
     David.
Hauled straw into the barn and fixed the hen house.
Chored about, and hauled manure for two days.
Took Mrs. C. a-visiting, and kept house myself.
Went to William Miller's on a bull trade.

December:
Went back to William Miller's with a bull.
Hauled manure and wood for two days.
Fixed about the house, wife went to town.
Caught a severe cold with choring and fussing around the
     house and barn trying to keep the animals warm.
Killed a wildcat and skinned a sheep that drowned in the well.
Spent four days cutting wood.
Went to Carimona for a grist of meal.

A Year's Expenses: 1880

Jan. 8th.	Paid for lumber & nails	$ 3.75
Jan. 20th.	Paid for lumber & nails	$ 1.72
Jan. 20th.	Paid for taxes, bits, postage, matches	$ 2.50
Jan. 20th.	Paid for old debts	$10.15
Jan. 24th.	Paid for beef & crackers	$.45
Jan. 24th.	Paid for Wakefield's Cough Syrup	$.50
Jan. 31st.	Paid for lumber & nails	$2.50
Jan. 31st.	Paid for groceries	$.53
Feb. 13th.	Paid for flannel, lath & nails	$3.43
Feb. 13th.	Paid for debt to Wm. Thatcher	$.75
Feb. 23rd.	Paid Dr. Jones for medical attendance	$3.50
Mar. 1st.	Paid for groceries & etc.	$1.43
Apr. 12th.	Paid for dry goods	$1.38
Apr. 13th.	Paid for groceries & etc.	$.80
Apr. 29th.	Paid for hardware, groceries, etc.	$1.87
May 8th.	Paid for groceries	$.75
May 28th.	Paid for coffee & sugar	$.77
June 3rd.	Paid for hat	$.30
June 17th.	Paid for groceries, shoes for Myrtie	$1.83
June 17th.	Paid for sheep & med.	$2.00
June 29th.	Paid for goods at Wheeler's	$8.55
June 29th.	Paid for goods at Benjamin's	$2.31
June 29th.	Paid for goods at Doughartie's	$.50
Sept. 13th.	Paid for harvest labor	$13.37
Sept. 19th.	Paid for boots	$6.50
Sept. 19th.	Paid for dry goods & basket	$3.22
Sept. 22nd.	Paid for horse book & ball-on knife	$.70
Sept. 28th.	Paid for lumber & nails	$5.55
Sept. 28th.	Paid for dry goods, etc.	$1.70
Oct. 6th.	Paid for hardware	$2.55

Oct. 7th.	Paid for flour	$.75
Oct. 23rd.	Paid for kerosene & barrel	$.63
Nov. 15th.	Paid for books, lumber & med.	$2.65
Nov. 19th.	Paid debts amounting to	$7.45
Nov. 19th.	Paid for school books	$2.03
Dec. 3rd.	Gave wife	$4.20
Dec. 18th.	Spent	$.72

Total expenses recorded in 1880   **$104.29**

A Year's Revenue: 1880

Jan. 8th.	Received for butter	$1.25
Jan. 20th.	Received for sheep	$2.50
Jan. 20th.	Received for wheat	$14.00
Jan. 31st.	Received for poultry & eggs	$1.53
Mar. 15th.	Received for eggs, etc.	$1.35
Apr. 12th.	Received for produce	$1.43
Apr. 13th.	Received for produce	$.80
Apr. 16th.	Received for moving a man	$1.00
Apr. 29th.	Received for butter & eggs	$1.76
May 28th.	Received for butter & eggs	$.77
June 3rd.	Received for butter & eggs	$.30
June 17th.	Received for butter, eggs & sheep	$3.83
June 29th.	Received for wool	$10.26
Sept. 13th.	Received for colts	$23.37
Sept. 22nd.	Received for colts	$11.00
Sept. 30th.	Received for butter	$.55
Oct. 23rd.	Received for butter	$.40
Nov. 15th.	Received for stock	$2.25
Nov. 15th.	Received for butter	$1.12
Nov. 19th.	Received for pork	$10.60
Dec. 3rd.	Received for work	$ 3.75
Dec. 18th.	Received for butter	$.37
Dec. 24th.	Exchanged poultry for merchandise	$1.45
Dec. 24th.	Paid Olsen for plastering in honey	$1.25

Total revenue recorded in 1880   **$96.89**

Fred's brother, J. Marshall
Cummings, c. 1863.

Fred's brother, Henry J.
Cummings, c. 1863.

Lyman and Mary Howe, Rose's parents (date unknown).

Left — Rose and Fred
Cummings, c. 1870-80 (?).

Below — Second class to graduate
Preston High School (1892): Louise
Baker, Lucy & Stella Gray, Jennie
Taylor (Warren's bride).

Kenneth Cummings, c. 1900.

Elvyn and Myrna Cummings, c. 1902 (?).

Waukokee School, March 1910, Warren Cummings, teacher.
Students, L-R: Vinton Cummings (next to Warren), Elvyn Cummings, Alfred
Dahl, Pete Vaalemon, Crystal Rose, Beah Dahl, Tillie Peterson, Alice Miner,
Verna Mathews, Viola Dahl, Mary Isaacson, Mabel Ballard. Front row: Kenneth
Cummings, Sidney Miner, Harold Rose, Ida Vaalemon.

Rose and Fred Cummings, c. 1910.

B. Franklin Cummings (Fred's brother) with Rose (R)
and her sisters: Samantha Gould (L) and Mary Burns
(bottom), c. 1918 (?).

Fred Cummings in the carriage with Nancy &
Fanny mares (date unknown).

Landscape of the bottom land,
view from the house
(date unknown).

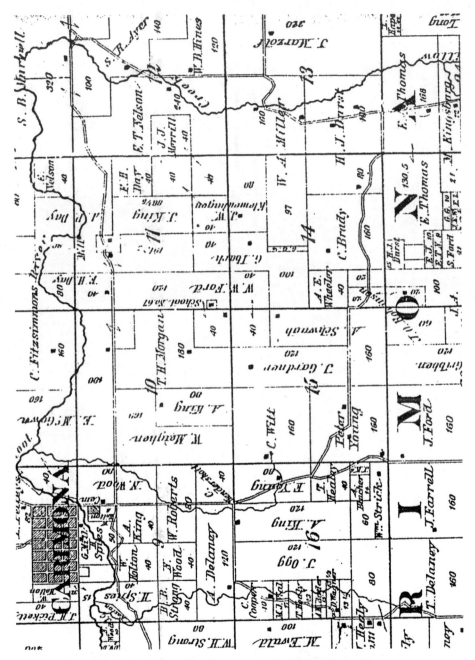

Section of the 1878 plat map of Carimona Township.

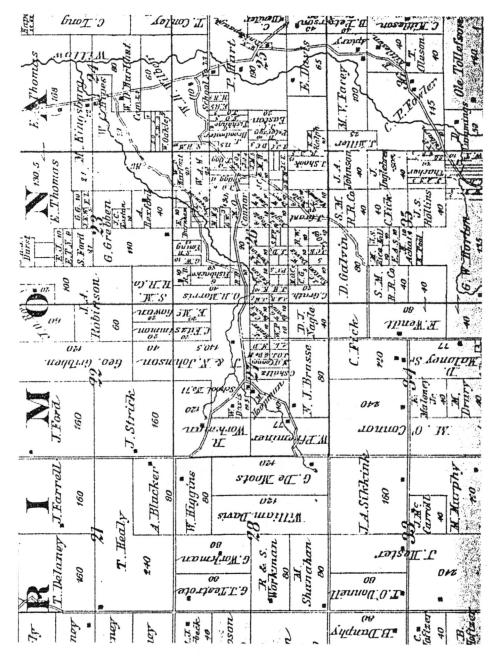

C. P. Fowler's and David Cummings's farms are in the upper right-hand corner.

The Shrunken Soul

If you have a soul, William Miller,
What kind of a thing must it be?
It's as hard a job to find it, I reckon,
As to capture an Arkansas flea.

If you have a soul, William Miller,
Say, where do you keep it at night?
It must smother, I'm sure, in your body,
Your hide is so beggarly tight.

You must leave this world some day, Bill Miller,
Say, where shall we look for you then?
A course of deception and lying
Points straight to the Devil's bullpen.

If you have a soul, William Miller,
It must be so daintily small,
That 'twould surfeit upon a flea's liver
Or could dance on the point of an awl.

I think of the quadruped kingdom,
Whenever your body I see,
But my thoughts are in constant confusion
'Twixt a thoroughbred bull and a flea.

I am not blaming you now for your stature,
Though your legs are most fearfully wry,
But your miserable, miserly nature
Puts to shame even the beast of the sty.

If you have a soul, William Miller,
Oh say, does it ever dare chide,
Does it ever dare trouble your slumbers
When you've robbed a poor cripple or lied?

I shall think of you often, Bill Miller,
A hypocrite, liar and thief,
A sheepskin upon a wolf's body,
Expresses my feelings in brief.

—1880

Jonah's Whale

The big fish eat the little ones,
This seems to be a rule,
But if he doesn't squirm around
The little fish's a fool.

Down in the whale's capacious maw,
The good old Jonah prayed,
The frightened fish with terror saw
The sad mistake he'd made.

The tortured fish thus preyed upon
Began to retch and heave,
He turned his body upside down
His stomach to relieve.

He twisted, dove, and turned about,
Then stood upon his tail,
He churned the sea, did blow and spout,
But all of no avail.

Moved by an impulse from within,
He struck out toward the land
And Jonah, meek and tired of sin,
Was cast upon the sand.

To those who wish to swallow me,
I've this advice to give,
A warning take from Jonah's whale,
And let your neighbors live.

For though I'm small and weak, I own,
And easy taken in,
I've such a habit when I'm down,
Of thrashing 'round like sin.

—1880

Death's Harvest

Fred experiences what we would today call a bout of depression. He had been cheated on a livestock deal, he is in debt, he becomes the caretaker of his sister's children, and he learns that one of his dearest childhood friends is ill to the point of death. It is no wonder that he feels overloaded and, as usual, he expresses himself through poetry.

Well, Mr. Thatcher relieved me of my heifer and a calf, and by so doing, reduced my debt to about $12. I didn't like that way of getting relief, but it was only the second act in the Bill Miller play. I entered into an agreement to purchase a bull from a neighbor, then found that I got a "bum steer." He refused to take the beast back and held me to my note even though I had not received the goods I expected. It was hard to bear down, but William should remember who said "vengeance is mine." I drew a curtain over the scene and waited patiently for the Good Lord to dispose of the case as it seemed best. I should make a botch job if I undertook it, I was sure.

We had a real blizzard and twenty-two below in mid-January of 1881. Frank came down and said he had worked his passage bunting snow. The railroad was constantly filling up with snow, which made traveling by rail very unpleasant just then. The settlers on the frontier in Minnesota suffered intensely from cold and want on account of the snow blockade on the railroad. Railway accidents and avalanches were numerous in the west, east, and everyplace where there was a road or a mountain steep enough for snow to slide down. Vague reports came to us that death from cold and hunger was staring the people of the western prairies in the face, but just the extent of their suffering and destitution could not be ascertained until the winter opened up.

The bluebird came March 27th, but the snow went off slowly. Frank wrote from Cottonwood County, saying that we had heard

137

nothing untrue in regard to their winter and that the reports of suffering and death were no exaggeration. Brother Marshall died that April in New Hampshire of lung consumption, after a sickness of two years. I had received a letter from him in February in which he said that he had resigned himself to whatever the future had in store for him and that he lived in hopes of a better state of things beyond this life for the whole human race. I had not seen him in thirty years, since I had moved to Minnesota. He was one of four brothers who served in the great Civil War.

My brother Moses went to New York state to visit his mother-in-law about this time. He decided to stay there, and I only saw him once more after that, neither do I know anything of his life out there. I remember his good qualities and forget his defects. He was endowed by nature with talents far surpassing my own. Some of them he cultivated and some he neglected for pleasure or worldly gain.

The excitement on the farm in January of 1882 was twins. A cow brought forth two male calves to Uncle Fowler. This year marked the first trouble with my eyesight. I had pain after reading a few minutes, especially at night. So I consulted Dr. Jones, who said glasses would not relieve me, that I was overworking my eyes and must stop reading when they began to complain. However, I got a cheap pair of "specs" at the drug store and found to my joy that they *did* relieve the pressure and pain. But I only used them when necessary and, as my sight was good yet, I didn't find it necessary to wear glasses continuously. So, of course, I questioned the need of it in a majority of spectacle-wearing victims, calling it one of the bad habits of weak humanity. But, I found they were a benefit to my eyes and used them as needed thereafter.

I heard a bluebird the 14th of February, a full month ahead of most other years. March began very warm and I put my bees out— fourteen colonies, good and strong and as heavy as I could lug. I hauled up summer fuel on bare ground with a sled. There was very little sleighing that winter, and much mud. School closed and the youngsters were at home and underfoot with their nonsense and surprising wisdom.

On Easter Sunday the young ones were cooking eggs. The weather kept disgustingly wet and oh, such mud, we missed our week's mail due to the mud. My winter wheat was green, we had six male calves, yet spring moved on slowly and the ground stayed wet and heavy. I had not sown any seed, but managed to plow eleven acres.

We had frost on June 11th and a slight shock of an earthquake the next morning. I dragged my corn ground over almost twice, but quit due to a big rain storm with wind and lightning. I don't thrive on wind, and that accounted for my being entirely satisfied when it quit.

Our little neighbor, Florence Colburn, went to Oregon and wrote back that she decided to marry in June. Sister Kate came for a visit with Ella and baby Ada.

We were building a buttery in the east end of our kitchen. It had cost $15.35 so far, yet it was unfinished. Eight gallons of honey made us glad. I now had nineteen colonies of bees. Farmers were working into winter grain. Most of them sowed rye and many risked up to twenty-five acres on wheat. This was the second or third year we raised winter wheat and rye, with considerable success at first, but it seemed by this time to run out into weeds and chess. I was not satisfied with the experiment, and decided not to try it again.

Warren was reading *Uncle Tom's Cabin*. He also began to saw a cat gut with horsehair (I suppose some would say play the violin), quite satisfactorily to himself and to some others as young and unknowing as himself. I contracted to have Warren take music lessons in both the organ and violin at the rate of $8 for twenty lessons. I was praying for good health and good luck the next year, for Warren was teasing for an organ, which would be at least $65 more.

There was a most splendid, brilliant, and beautiful comet to be seen in the early morning toward the middle of October; it rose about three in the morning and was almost in the sun's path. This was when I first heard the news that my friend and classmate of 1864, Helen Powers Moore, had been sick for many months and was slowly dying at Isinours. This was most distressing news, as Helen had been a dear soul to me since my youth. The effect of the news on me was a dream of hearing her voice as of yore, singing as clear and sweet as ever I heard it. And it seemed to ring in my ears long after I awoke, a familiar hymn tune, but words I never heard before. The first chorus was as follows:

> Don't you hear the holy songsters
> As they beckon me to come?
> "Zion's gates shall open for you
> Welcome sister, welcome home."

Aunt Mary was here, and she prepared to marry. She became Mrs. George Burns, and they left to settle in Waubay, South Dakota. Kate sent word that I must go to the station to get her girls. This was how VanSickle sent me and Uncle Fowler his family to support, five children.

Spending 75¢ more for my house made $22.90 that year, besides a vast amount of labor, but such is life. I didn't want to die, but the truth is, I don't believe anyone ever benefitted by my living, nor did I know that God ever cared whether I was a man or a mouse. The problem of life was too much for me some nights; my back ached and no one loved me. I made a bad job of things again, and was head over heels in debt: $18.35 to Conkey, and $4.60 to Watson. The cares of life crowded heavily upon me. I felt as though a little more would discourage me completely. I suppose that it was the same trouble that makes a free horse balk, viz., overloading. If I could only balk, I believe I would. And if I only knew that God required me to pull through for some purpose, I would bear up more cheerfully.

My health was not very encouraging at the time and, as usual on such occasions, I was tired of earth's oppressive, crooked way and wished many things in my ignorance, and all to no avail. Here I was and here must I remain and toil, knowing only one sure thing: there would be rest by and by, perhaps oblivion. One thing more I knew, nature had no favorites, none were exempt. Those we loved most and best faded and withered with the rest.

Reflections

I'm alone in this cheerless, heartless world,
No sympathizing heartthrob do I know.
I've not a friend on earth to share my woe,
Or feel the joy stern fate permits to flit
Like transient specters o'er my lonely path.
True, there are those who by my fireside love to sit,
And those who ask about my state of health,
And kindly remedies appeasing nature's wrath
For some offense that this poor erring body doth commit.
And there it ends — no never-dying world of wealth
Can take the place of never-dying unison —
That answering chord that vibrates through the soul
Of all around, like a well-tuned instrument whose chords
Will answer back the sound of each
And recognize themselves as parts of one great whole
Who cannot live apart, but run
Together like the rays of heavenly lustre
Glimmering from Nature's life-inspiring sun.

The first part of 1883 brought us a very severe spell of weather and a blizzard. I kept busy doing chores: nine cattle, seven calves, forty sheep, five hogs, forty chickens and three horses to care for and I had to keep their stables warm, which left me little time for anything else besides the necessary house chores such as hauling wood for myself and Uncle Fowler.

The cold wave lasted one solid month of zero and all the way down to thirty-eight below. I worked hard to keep the stock warm, fed and watered, and no lives were lost. The program was: warmed oats twice a day for the calves, corn prepared like the oats for hogs and cattle twice a day, and the hens were treated once a day to roasted corn.

By mid-February we had a thaw, a thundershower, and a freeze up. I hauled three cords of wood to town for C. P. Fowler and planned to take down three-fourths of a cord more. My liver complained very much, and I didn't know but that Dr. Jones would have to look after me. That's the way the money went.

Helen Moore died February 28th, and they brought her poor, pale face to the schoolhouse where the funeral services were held, and then laid her to rest in the Waukokee cemetery on Friday, March 2nd. Thus passed another gentle, lovely soul as I ever knew. One of the best friends of my childhood and youth, and the thoughts of those days come as near making me cry as anything that had happened for years. I loved that gentle soul with a never-dying love, and now my only hope of perfect bliss in the life to come was a meeting with Helen Powers, as I used to know her, and to hear her sing that sweet hymn once more.

Sunset Musings

We shall know each other there;
Dear wife, please leave me to myself tonight
Nor watch me with such nervous fear —
'Tis sorrow and not pain — I'm not sick again,
No, I'm not sick again.
But oh, I've heard a thing that made me weep to hear,
Long years ago, you do not know
How sweet the memory is to my poor soul today,
I had a friend, a classmate, dear sweet-tempered soul
And I loved and even worshipped her, dear wife, don't call me
　　fool.
For 'twas no foolish, moonstruck love, but such as angels feel,
And now a messenger has told me she is dead.
Oh, can it be, and is friend Helen gone?
I call her friend because she was another's wife,
And I did never envy him his prize, but how
I can but think of her as standing by my side,
With chalk and book in hand as we strove to win
What none might take away: a scholar's crown.
She won the admiration too, of all the good and pure in heart,
And I forever must adore and long to see that face again,
But all too well I know such wish for mortal man is vain.

This little poem written to her memory might give anyone an idea of my sorrow at the news of death's harvest, and the sense of loneliness that comes over me when I realize that those I loved most and best in my boyhood days will never speak or sing again on earth. It was a sad pleasure for me to visit her little grave occasionally with a bunch of flowers to leave there to wilt in a few hours, to pull weeds away, or to readjust the headstone that would settle and tip again as time changed the mound.

The city of Rochester was rent asunder by a cyclone that fall, which was only one of many such terrible things that swept over different parts of the United States that year. People began to feel as though old earth was an uncertain place to stay. Astronomers thought they knew the cause of these twisters. Full particulars from Rochester fixed the number of killed by that storm at thirty-five, and one hundred wounded. Then we learned that a big earthquake on the island of Java made terrible, terrible havoc with humanity: 75,000 lives lost in a day, with attendant horrors in proportion. Men's hearts failed them from fear, showers of red-hot rocks fell a distance of forty-five miles from the volcanoes.

November was a terror to threshers. The wind commenced to blow "Hail Columbia" from the northwest and the mercury went down to zero and held its own. I did not vote that year, but the Republicans had it in our state.

We had a queer-looking sky all the month of December. In the early morning and at dark in the evening the sky turned as red as blood, with no clouds in the sky at the time; and sometimes when it was cloudy the bright light of the sky beyond seemed to penetrate them. All kinds of theories were afloat in regard to the phenomenon in which superstition and fear played a big part. We concluded it was caused by the Java earthquake and the dust from the volcano. This I supposed to be the correct theory, strange as it may appear.

— o —

Assassin's Bullets

The newspapers reported that the poor old czar, or emperor, of Russia was murdered in a most shocking and brutal manner. This was one of a series of attempts to murder this noble old monarch — no less than five assaults had been made on him by the nihilists with dagger, pistol, and finally, by infernal machines. Just what the nihilists were, I could not say, but supposed them to be a set of men banded together and sworn to do away with all government, or at least, the present form of government. It seemed also that these men cast lots, and upon whom the lot fell, fell the terrible alternative to attempt the deed or die by the hands of his fellow conspirators.

On July 2, 1881 news of a not very welcome nature reached us by telegraph, it said that President Garfield had been murdered. Then we learned that President Garfield was still alive, but with a slim chance of recovering. He was shot in the back by a half-witted wretch who wanted to become noted for some big thing. The Fourth of July was a day of national mourning in consequence of this foul crime.

Reports dated two weeks later said the president was still alive and would probably recover, although the bullet could not be found as yet. Mrs. Garfield bore the shock with great fortitude. It was claimed that the president's recovery was entirely due to his own indomitable will and fearless determination to master the situation, regardless of his terrible suffering and seemingly hopeless condition. Never did a man receive such care and never were the American people so anxious and so deeply interested in a man's distress. Everything that could be devised for his comfort was provided. Four doctors tended him, one in constant attendance day and night.

Our president held out manfully. The fool that shot him said he wished he had been idle. Most of us wished he had frozen in a snowbank the previous winter. The doctors thought they had succeeded in finding the ball that was shot into the president's back. Some Yankee genius invented a machine by which the exact locality of a ball could be determined, and the machine said the bullet was in the walls of the abdomen about five inches below the navel. Mr. Garfield kept improving, although the surgeons had to open the wound in order to keep it running.

By mid-August we learned our poor president was having a difficult time. It didn't seem possible that his strength could hold out much longer and news of his death would not have surprised anyone, yet we clung to the fond hope that his life would be spared. The president was greatly reduced, having lost eighty pounds of flesh since the tragedy.

In early September they took the suffering man to the seashore in the same tender, painstaking way that was pursued throughout his convalescence. Removing the president from Washington to Long Branch without accident or incident gave the people new hope. His condition remained about the same, but doubt and uncertainty enshrouded the nation. Now that we were about to lose him, we could appreciate his worth, and we prayed to God to spare his life.

A double guard was placed around the assassin, not to prevent his escape, but to prevent his being lynched. The wretch was in constant fear. An old soldier forgot himself, attempted to shoot him, and missed his mark by four inches. The rash soldier was immediately arrested.

The last act in the tragedy passed as James A. Garfield died on September 19, 1881, wasted and worn to a mere skeleton—he died from sheer exhaustion. His powerful frame, vigorous constitution, and undaunted soul bore up under the fatal stroke eleven weeks and two days, until almost every part of his body was attacked with a special trouble, when neuralgia of the heart closed the scene and finished his laborious struggle with death. President Garfield's funeral was observed all over the Union, bells tolled and sermons with other appropriate services were the order of the day.

Our Heavenly Father worketh well
Although his work and will be death,
We'll trust his loving kindness still;
We'll mourn our loss, but sure God's will
Hath wrought the change, withheld the breath.

So dies the brave, the noble and the true;
The victim of a coward, knave and fiend.
It must have been God's will, we know,
To Him we bow, on Him we've ever leaned;
Our noblest Roman has not screamed
From this most dread and undeserved blow.

By December the miserable devil that shot President Garfield was having his trial—insanity was his plea. He hoped to get excused on the ground of "*non compos mentis.*" They were having a wild time with the poor fool. A little halter exercise might have saved the government thousands of dollars. Here I wish to emphasize my opposition to capital punishment, but it was the law, and I was decidedly in favor of giving this wretch the full benefit of it.

Garfield's assassin was declared guilty of murder after a two-month trial. The defense claimed insanity, but the jury could not see it. If they had pled devilish depravity I guess they would have won the case.

Sergeant Mason, the soldier who refused to guard the devil that shot Garfield and took a random shot at him was sent to prison for eight years. Efforts were being made to induce the president to pardon him. Surely, his crime was nothing more than refusing to obey orders.

President Arthur didn't do some things as we hoped he would, though he was a very acceptable officer. However, taking our law-making power as a whole, they did very many ill-advised things such as placing Grant on the "retired list" with an annual pension of $20,000 and restoring an old reprobate general to rank who was cashiered in 1863 for willful disobedience in the Battle of Bull Run. The mean part of all this was the money given the miserable drones in the form of back pay. Ulysses S. Grant was a good soldier, he received ample pay, and for president he received twice the salary of his predecessors. This piling honors and money upon him was downright nonsense, and an insult to every Union soldier.

I wrote "Blizzard Joe" to remind our pension-granting power that there were others who had served our country and deserved a help. It was very satisfactory to note that, although my poem never went outside the house where it was born, Congress finally did pension the old boys who did the work quite liberally at $12 per month.

We heard that H. W. Longfellow was dead, "America's famous poet;" he was seventy-five years old and well-beloved by all. Like Bryant, he was a special friend of my childhood. "The Village Blacksmith" was one of the first poems I ever tried to commit to memory. The notorious Jesse James died, too, but with a ball through his head. Another bandit met the doom he richly deserved so long. Mrs. Lincoln, wife of that noble hero of 1861-65 died about then. She had been partially deranged since Mr. Lincoln was murdered.

— o —

Blizzard Joe

Ye senators and congressmen, come listen to the prayer
Of a maimed and crippled citizen who nothing asks unfair.
My name is rarely mentioned, none but my neighbors know
Why I, a one-armed soldier boy, am nicknamed Blizzard Joe.

For I have been a veteran, I've heard the bugle bray,
But that's not why I'm crippled up, a one-armed man today.
No, I escaped the rebel lead and rebel prison pens
To meet a more relentless foe upon the western plains.

Yes, when the war was over and we were mustered out,
No man in all my regiment was counted quite so stout.
With joy I laid my musket by, and took my saw and plane
And laboring at some building, fought my battles o'er again.
I told my warlike stories to eager, listening ears:
The camp, the march the desperate charge, the panic's crazy
    fears.

In peace and joy the years rolled on, my labors all were blest,
And with my wife and babies, one day I started west;
There taking up a homestead, ten miles or more from town,
I made my trade support me when the grasshoppers came
    down.
One night, while tramping homeward from a two-week lucky
    job,
I fell to thinking o'er the past, oh, how my heart did throb.
For I was gay and merry, though 'twas not with poison wine,
I had finished up a contract with the railroad line.
There was money in my pocket and I carried on my back,
Some comforts for the children in a little paper sack.
The autumn clouds hung heavy but 'twas warm and early yet,
And, though walking on quite briskly, I thought there was no
    fret;
I was thinking of my supper and a faultless cup of tea,
While the children drove us crazy with their rattletrap and
    glee.

Although a mile or more away, our cabin was in sight,
And I could see, though scarcely dark, that Nellie had a light.
When, suddenly, an awful sound burst on my startled ear,
The roar of rebel canon never gave me such a scare,
"A storm in the air," I said, but bravely faced the blast,
Though the shot and shell from Keenesaw ne'er came so thick
    and fast.

Oh, desperate was the struggle, none but a squatter knows,
How quick the weather changes or how cold a blizzard blows.
My clothing, made for summer, seemed like a fisher's net
And I battled with the storm until every thread was wet.
Then I turned, gave up the contest, and retreated with the
    storm
Yet by no means prepared to yield to death in any form.

Bewildered, lost, half-frozen, I knew not where to go
But in hopes of finding somewhere, I wandered to and fro.
Then for my helpless children, I prayed in my despair,
And I thought with desperate longing of that coat I used to
    wear;
But not a sign of pity came from out the stormy sky,
My poor limbs refused to bear me, I was left alone to die.

Now again, my ears seemed greeted with the battle's awful
    roar,
And heard our captain urging us to rally just once more:
Chickamauga, Lookout Mountain and the long tramp to the
    sea,
Came to mind as plainly as it ever came to me.

Then came a dread sensation, a horrid, wracking pain,
It seemed as though a score of darts were driven through my
    brain;
And sounds I heard of wailing, a stifled sobbing low,
Gave me the startling fancy — I'm in that world of woe.
Then all the warning sermons and all I'd heard or read,

However void of reason now came trooping through my head.
Then, how Nellie and the babies would fancy me in Heaven,
"Pa was so kind to us we know his sins must be forgiven."
And I thought that they would search for me among the
    millions there,
Oh my soul was wrung with anguish and I groaned in wild
    despair.
How long I lay thus, weltering, I won't pretend to tell,
But I make this free confession: I have seen enough of hell.

One day I woke to consciousness, weak, sick, and sad enough,
For I was minus half an arm, someone had cut it off.
Then they brought the sack of trinkets—they had found it by
    my side;
And they said my frozen fingers, its mouth securely tied,
And the chances are they tell me, although nobody knows,
If I had thought to drop them, I shouldn't have been froze.

Do you grow impatient, statesmen, waiting for my tale to end,
And to know what 'tis I'm asking of a poor ex-soldier's friend?
Well, 'tis simply that you'll legislate to help the nation's poor
Who stood within the withering ranks and faced the iron
    shower;
Moving as ordered, right or left, amid the storm of death
Trusting to those above us, though panting oft for breath.
Though we escaped uninjured in either wind or limb
Our general did, and you saw fit to doubly pension him.
But we ask no lordly legacy to make us rich and great
Just keep the poor, maimed soldiers clear of starvation's gate.
Where you give Grant 1,000, give us a single five,
And rest assured that no complaint will e'er be heard from
    those alive.

While we give to him the glory, a little praise we take,
We were slaughtered by the thousands, by many a sad
    mistake.
If to him alone be glory, he too must bear the guilt,

With the bones of slaughtered thousands, his monument is
    built.
'Tis a pity that one veteran should ever be passed by
When sad misfortune lays him up to beg, alas, or die.
While all the land is running o'er, the land he fought to save,
With wealth and great prosperity, oh play us fair we crave—
Give every worthy soldier enough to help him rise
When sickness, sore or accident cuts off his scant supply.

—1882

Bootless Victory

Fred mentions a letter from Rev. George Mooar, who was "searching for Cummings." In 1903 Mooar published what has become one of the primary reference books on the genealogy and descendants of Isaac Cummings, a settler in Ipswich (Topsfield), Massachussetts, circa 1635. The Cummings Memorial is fairly comprehensive through eight generations, up to a few years before it was published. Even so, the book is not totally accurate, as it lists Fred's brother Frank's first wife as Jeanette Borden instead of Bowden, and spells Myrtle's first name as Phebe instead of Phoebe. These two errors are repeated in Cummings Genealogy by A. O. Cummins, published in 1904, but he also credits a fifth child to Fred and Rose, listing Myrtle and Phebe M. as two children, not realizing they are the same person.

Quite a curiosity turned up in early 1884 in the form of a letter from one Reverend George Mooar of Oakland, California — he was searching for Cummings. He asked for a record of our family as far as I could give it. He gave me the information that the Cummings family could be traced back as far as the year A.D. 1620 or thereabouts. In this country, most were located in New Hampshire; Scotch–Irish their origin, whatever that meant. I asked a Scotch man what it meant, and he said, "An Irish man ashamed of his nativity." So now, when asked about my origin, I say "Irish," for I am not ashamed of Old Ireland.

Brother Frank was here, and stayed about three weeks. I bought his buffalo coat for $8. I sold $15 worth of honey to I. N. Howe and $18.62 worth of beef, so that I almost swung clear of debt at that time. I only owed $4 on my coat and a note of $5 to Conkey Brothers.

Our creek froze dry and remained so for weeks, much to our discomfort, as we had to drive stock to Facey's farm for water from his spring. But, elsewhere, there was terrible suffering from wind and flood. The Ohio River overflowed its banks and swept everything before it, and several of the Southern states were visited by a terrible cyclone, which made the heart sick, for the word meant death the most fearful. It meant that cities were rent into fragments in the twinkling of an eye, its panic-stricken inhabitants scattered, wounded, and dying in every direction.

March 20th produced the bluebird and brought springtime again. Almost all of our snow went to pieces without any big flood. I had been taking medicine for a week for rheumatism, neuralgia, and general cussedness. It was the same old story: lame side, sore spot, and blister on same makes sore spot sorer; greased rag on blister, bottle dope inside, and one dollar cash out of pocket. This last was a problem since there was only 50¢ in the pocket to begin with.

I put my bees out—fifteen hives—and thought they did not look very promising. I took Grandpa Howe to Fountain after a visit here of two months. The turtle dove was singing in the woods Easter morning, and people commenced farming.

Wolves gave our sheep a terrible hack, killing and downing five. Wolf hunting went on vigorously, but whereas there were many seekers in the field, there were few that got the wolflets. Such is life, there aren't enough to go round when we hunt them, but when they searched for us we were outnumbered and out-generalled, out-flanked and out-of-pocket. These creatures caused us no end of annoyance by their depredations and cunning tricks that made catching them a hopeless task.

I made another pass with the wolf family, and claimed another bootless victory. I caught a big brute and took him to the county auditor for the $7 bounty, but I failed to get my reward because he looked "too dogish." I maintained he was a wolf just the same, but I could only content myself by saying I'll try again.

My new cyclone cellar in the side hill above the house was all done but the finishing. I estimated the cost of it at $30. It was ten by twelve feet, and the inside had a solid sand rock floor. It was covered with about eighteen inches of dirt upon a stone roof, which rested upon stout timbers that were about twelve inches apart.

A storm of much rain in August knocked my oats down badly and raised the water over some grass that I was going to make hay of the next day. I lost six or eight loads by it. The express brought a Scotch collie pup to my address, at a price of $6.85. This would do for me a lifetime, but some here thought it a good investment. One thing certain, the thing was smart and as mischievous as a monkey.

The above expense, with $14 for a tombstone for little Reuben and $2 for apple trees, covered a list of miscellaneous expenditures that took longer to pay than they did to incur. But I had to, the baby had laid under a rough limestone slab nearly ten years. This might be all right for some, but tombstones are marks of civilization as much as anything, and it seemed a necessity to me. Mr. Powers put up the tombstone in September, and it was a thing of beauty.

Little Reuben's Grave

'Tis but a mound, a simple mound,
Which does a baby's ashes hide.
And there are dozens more beside
In this, a people's holy ground
   Where loved ones dear
A final resting place have found.
Why is it that the tears will flow
While bending o'er this mound of sod?
Oh, why do I worship this cold clod,
When it reminds me of heartrending woe:
   A mother's grief,
And the last agonizing throe?

Oft in the stillness of the night,
When slumber rests on all around,
My waking thoughts fly to this mound.
Once more I see those eyes so bright
Of that sweet child with heart so light,
   And tears will flow
To think upon death's withering blight.

Yet something seems to chide my grief,
Methinks it is an angel's voice,
Which says it is Our Father's choice
To give this little soul relief
From this cold world where joys are brief
                    And take it home —
God's will be done, oh help my unbelief.

We threshed in September, and I had 192 bushels wheat and 391 of oats. The following was my debt account as of October:

To Conkey Brothers, $7 for one suit clothes for myself; $2.25 for
    groceries
To Watson for boots and shoes, $7.65
To C. Peterson, on horse rake, $3
To Tom Dougherty, on plow, $11.50

Election day was past, and for another time we engaged in a test of strength politically. Four prominent men were before the people for election and one of them was chosen president. I voted for Blaine and Logan while, could I have had my choice, I might have voted for the Prohibition man. But while prohibition is a principle dear to me, I couldn't help thinking that there were other things involved, which would be dangerous to ignore or risk in Democratic hands for the sake of rebuking the Republican party for some of its errors. But, it was Democratic reform that won the day. We knew we had to stand four years of Democrats, then we would see about it. The Republican party ran its best men and were beaten by a very small majority. The man hoisted into power by this chance was an inexperienced one, but we hoped he was honest. Republicans said we could expect rum and rebellion to triumph for a season.

We had a short visit from both brothers Frank and Moses that year. Frank came from the west part of this state, and Moses from New York, and I supposed it to be the last time I would see either of them. Moses thought me a bad man for getting entangled in someone else's trouble. Some of my neighbors got into a foolish scrap and, by telling me half a story, enlisted my sympathy and engaged my help, which I gave in full measure. In fact, I don't believe in half-hearted support of any cause, so I tangled myself up in another man's war in a most ridiculous shape, before I became aware of the real facts. Then I set about to draw off and save my honor but, have to confess, like the Scotch soldier, "There were no' a very great amount of honor in it." Two or three poems were the outcome of this episode, but not to my credit.

Snow lay about two feet deep in the woods through January 1885, but February brought us a fine day, which meant the same for a month to come, so said the prophets and groundhogs. I hauled wood for Uncle Fowler, as he was unable to get his own that year. The rest of my time was spent working for the brutes at the barn and the folks at the house.

The weather prophets were wrong. Cold, awful cold, was the song in the air through the next month. The mercury gauged from ten to twenty below zero every morning, and hardly got above during the day with a north wind blowing. Grover Cleveland took the presidential chair March 4th, and said that he would do the best he could.

I got a lot of wood hauled up and then turned to getting the corn that wintered in the shock out, and estimated that I had about thirty bushels of ears left. The bluebird came March 26th, also a bald eagle — these birds were rarely seen in our vicinity. I made some sap troughs and Warren kept busy in the maple sugar business, but the weather was not very flattering to farmers and beekeepers. I came near losing all my bees and knew they would be in more danger before the rallying season came, so I prepared for the shock.

I may as well make a note of the fact right here that I was nearly thirty-nine years old and my teeth were all good but one. This one gave me trouble enough for a whole family at the time, and the future outlook was not pleasing to contemplate. The wisdom tooth was so annoying that I tried to get it out, but after two or three miserable attempts I concluded to bear it, as it utterly refused to go.

We learned about then that General Grant was dying of cancer. It was said he bore his suffering with a fortitude unparalleled in

history. We feared our well-beloved General Grant was probably
beyond the sorrow and pain and vexations that had been his
constant menace and torment for the past year. Harassed by financial failures threatening disgrace to his great name, broken in
health and in constant pain, his noble spirit had borne up under all
these calamities combined, and his character shone brighter as he
neared the end of life. The hero of the Appomattox was almost lost
sight of when compared with the dying hero of the day; no
complaint and hardly a groan escaped him, although we knew he
must be enduring untold agony. Surely, here was a picture for
painters and poets! General Grant lived on for several months, and
was gaining strength in spite of the dread destroyer at work at his
vitals.

I finished planting corn and potatoes by June. I had a very slim
team to work with that spring, which made me late with field
work. I went to James Bowden's and got the bees that I bought of
Frank Frasier. The bee business suffered badly, as I lost most all of
mine. Some did better than I and some worse, if that was possible.
My Scotch collie raised four pups that spring, and I tried to get $4 a
piece for them.

About an hour after writing in my journal on the 7th of June, a
quarter of my house roof was torn off by a tornado. I shall be apt to
remember the particulars of that fearful Sunday for my entire life. I
hate to re-live the terrors of the tornado's work, and so shall make
the story of my experience as brief as possible. The cyclone came
upon us about noon, tearing through the timber west of us, driving
us into the newly-made cellar. It took off part of our house roof
and scattered shingles along its path over the east hill, driving one
pine board through a small basswood, sweeping down many trees
on the hill, and then passing on to other fields. The cost of repairs
included two days' work on the house and a week's time for
Warren, repairing fence. I concluded to bear the ills I had on this
earth rather than fly.

Warren, who was fifteen, helped at home then. That month
was nice weather for fixing up corn. We went through the corn
three times, cultivated potatoes thoroughly, mowed weeds off
from eighteen acres of spring grain, fixed the barn up again, and

improved my cyclone cellar besides working on the highway mending fences and the like.

I don't know but that I made a mismove, buying a new mower, but could do no better anyway. Neighbor Martin Ogg and myself bought one on time for $63, with half to be paid that fall and the other half one year from then. We finished haying the last day of July and had thirty-three good loads of good hay in the barn. While mending fence, I found a very ugly looking rattlesnake measuring nearly four feet, having nine rattles.

General Grant went to his final rest at last, peacefully, quietly breathing out his life like a weary pilgrim gone to a long-desired destination. Mortal pen could hardly do justice to his memory and no man, however talented, could hope to pay a fitting tribute to this, the greatest of earth's heroes. And, perhaps the most wonderful accomplishment of his life was the completion of his book while suffering with the most painful disease that a mortal can have. No man ever did such a thing before, and probably never would again.

By August, grain was cut as fast as binders could be carried over the fields by three- to five-horse power. These machines were one of the wonders of the age. Harvesting machinery had advanced to the binder that year, which was in common use. Every other kind or contrivance for cutting was pronounced too slow and rejected at once and for all time. I, feeling too poor to pay the price, tried to hire one. But, as every farmer must cut his own first, I was driven to the dire extreme of hiring Jim Bowden's old East Off reaper to cut my fourteen acres of oats and bound them on the ground, as was done twenty-five years before, and considered right. I got the grain all right, but Martin Ogg, remarking on my style said, "I see you don't believe in procrastination" (meaning progression.) I thought this remark quite appropriate, and assured him I did *not* "believe in procrastination." Anyhow, I finished binding and shocking my grain after a hard week's work. My arms were very lame; my right arm had been lame a long time, and I felt thankful to have done as well as I did.

It was while riding the aforesaid reaper that I composed the verses "Funeral of General Grant." It was a national event, and every telegraph station was kept informed of the details as they occurred at the place of burial. Minute guns were fired in all the big cities and military posts.

Funeral of General Grant

There is a time for men to mourn
And in dumb sorrow bow the head,
The war-tried veteran, battle-worn,
In grief bewails the noble dead.

Nor can the soldier claim alone
A right the sable garb to wear,
A more-than-soldier now has gone,
Statesman and patriot claim an equal share.

For he was ours, each pang and groan
From north to south, from east to west,
Was felt and heard as 'twere our own,
All thought they loved the sufferer best.

Nature is king, and howe'er great
A mortal's mind or heart may be,
He finds that in his best estate
Life is but care and vanity.

— 0 —

Scattering of the Flock

Even though Fred does not write much about his children, he is distraught at the thought of them growing up and leaving home. He has created his own family, but now anticipates and dreads a time when they assert their independence.

Warren stayed with Dr. Love during the spring of 1886, and went to high school in Preston. Measles and mumps were going the rounds. We were not stricken yet, but hardly expected to escape. Warren came home in early April to help saw up our summer's wood while waiting for the reluctant snow to leave. Two weeks later, he had symptoms of mumps.

I commenced spring farming, and finished seeding in early May with about twelve acres of oats and five acres of barley. Warren dragged it while it was my turn to be down with the mumps. I was terribly sick for several days, much to my disgust, for I had thought mumps a trifle till then.

Harvest arrived by August after a long dry spell, making many long faces as usual and some sunny smiles. My crops were very good, wheat, barley and oats. Corn promised a good yield if we could get rain, but I feared potatoes would be few in a hill. The cattle were very troublesome on account of the drought as they searched for food in any field. I finally fenced off the meadow and put them there, enlarging their browse area by at least fourteen acres.

Lewis was in Dexter at his grandpa Howe's for two months that summer. I went for him and was treated to a prairie thunderstorm while returning from there, the heavens being lit up by a barn burning from lightning stroke.

I finished my potato digging and got thirty bushels of number one potatoes, and about seven bushels of little ones worth 50¢ per bushel. We had used ten bushels before this, making a total of forty bushels worth $20 for that year. The little ones went to the sheep.

I was making a crib for my corn and hauled much manure out that fall. I did a job of grading in the barnyard. The ground was so low next to the barn it was next to impossible to get there in wet weather.

Warren went to Grandpa's in Dexter to work in threshing. This year marked the beginning of a scattering in my little flock, which tore my hopes and heart into shreds. Disappointment followed every hope I ever entertained, until I gave up looking for anything else but separation from the children, and knew I must harden my heart to the inevitable.

Women were a-going to vote for county superintendent that year, as there was a lady candidate running against Brady of Preston. The women voted and, consequently, we had a Prohibition superintendent of schools that time.

I got us a good deal of trouble when I bought that mower on time the year before. I became the sole proprietor of that mower, but still owed $34.25 on it to Uncle Fowler. This made $57.25 worth of machinery he charged to me that year. Winter was here in early December, with twenty degrees below zero and fourteen inches of snow. A calf died, which was a Norwegian one I had bought of Father two months before. I planned to butcher the old cow by the first of January. I also had about twenty sheep at the time, and Uncle Fowler had another thirty-one, making a good-sized flock to feed.

Cold Christmas, cheap cattle, careless creatures, call it hard times. I had two pigs, four sheep and a cow all fattened but there was no market for them. A man offered me one cent per pound for my cow, which I declined.

Winter wore away in early March of 1887. Warren lived away and went to Preston high school again. Relatives came here for a visit, then Mrs. C. went to visit relatives, and Mr. C. was very sad, and grew sadder as the night approached.

He had a dream the other night
When all around was still,

He dreamed he had a letter
That came from Conkey's Mill.
My Darling Sir (the letter read),
I write you in a flurry
For I must write once more to Fred
To tell him not to hurry.

Ten dollars will fix Myrtle right,
Ten dollars more for me;
And I will meet you some dark night,
Together we will flee,
And we'll be off before the dawn
To where nobody knows.
'Till then, I'm your beloved fawn
And faithful firefly — Rose.

I heard a bluebird on the 10th of March, and saw one two days later. Mrs. C. came home on St. Patrick's Day in the evening. Warren got his first teacher's certificate from headquarters — following in his old father's footsteps.

After trying and failing times without number, we finally bagged a wolf and got our $7 reward, and the beauty of it was that we sheep-raisers got the entire benefit of the killing. Our cow, Molly, remembered us just as she had the last Easter, with a nice calf. The frogs and the whippoorwill tuned their pipes in the evening.

We had Scotch collie dogs for sale again that spring. We broke five acres of field on the hill — I was thinking of trying corn on it. We worked Billy, the four-year-old colt, some then and he went well.

Warren was helping me do the work that spring. Lewis and Myrtle went to school to Miss Baker. I planned to get a hive of bees from Jim Bowden to keep up the stock. My bees were played out due to entirely too dry weather and mismanagement. The dry spell continued, and scorching hot it was too. We were blessed with a fine shower at the end of May, the first we had in six weeks. People were looking very blue in consequence of the long, continued drought.

With Warren teaching a six-week term of school in the spring, Lewis became a necessary factor in our farming and stock raising industry becoming a herdsman, or cowboy, for it became absolutely necessary to watch the cows in the timber. He had half of each day allotted to this charge during the dry weather, until feed got better in the pastures. The west boundary line was surveyed out, and a line fence of barbed wire was put up on the farm in June for the first time since a white man lived here. This fence would outlast us all, or, at least the wire would.

Another month of dry weather passed, and hard times stared us in the face while we did the same to hard times and to each other. We had not had a famine in Fillmore County yet and, even then, we pointed with some degree of pride at the waving corn as the "rock of refuge." But that, too, threatened to fail if we had a continuation of that drought and the chintz bugs that were so plentiful. They went for the corn in dead earnest in July, but the clouds were coming to our relief, we hoped, with bountiful showers. We felt very much like a beleaguered garrison when help appears on the distant horizon. Will it come in time and succeed in overpowering the enemy?

It turned out that rain didn't do very much overpowering, but we got a shower two weeks later and another one a few days after that, which some of our thrifty farmers said would make some corn in spite of the bugs. Such steady, scorching sunshine, scarcely ever was seen. Steady, hot, dry weather continued all through July and was all we could see or think of, but some of us would have pretty fair corn. I paid for my harvest that year by cutting hay and thought myself lucky, as my oats were quite good.

After stacking my grain and doing other chores I took a layoff for five days in August to visit Dexter friends and found them better fixed there as regards pasturage and hay than we were. Much better, in fact, there appeared to be no lack of fodder there and for a distance of twelve or fifteen miles on this side, crops didn't seem to have suffered much on account of drought, either because their country stands it better or they had more rain, or both, perhaps. The prospect for work was so good that Warren went to try his fortune there for a time. While there, Warren found a church organ for cheap money—$18. I went after it with the team and brought it in. Fortune favored us and it was safely stowed away in our parlor. Warren's organ was nice to have around but, although we probably got the value out of it, I would never buy a second-hand instrument again with the history that old organ had.

Our weather underwent a marked change, and in August we had quite a good bit of rain and grass sprang up surprisingly. My bees bothered me trying to swarm, and I reduced their stores, taking about three quarts of buckwheat honey to pay for half a day of fooling with them. Besides that, I got eighteen or so stings thrown in gratis.

It rained all day long one day, but the thirsty ground took it all in and old Willow Creek was still very low. Our pasture was the last on the line to keep its water supply, which had always been reliable. The creek went dry below our place, and was dry down at Waukokee.

We drove some steers to town and saw many strange things at the county fair in late September. Part of the show consisted of a painless tooth-pulling exhibition, which looked very easy both for the puller and the pullee, but when the process was understood, most people would rather watch the show than be in it. The drug cocaine was the agent the "faker" used, and I would not trust my fate in a stranger's hands, especially such an odd-looking stranger as he was.

My own wisdom tooth that had troubled me for a while was almost covered up in the gum. Dr. Jones undertook the job of extracting it, but failed again after three desperate attempts, and the job was put off until later. So I kept my troublesome tooth for future torment.

We had a very severe storm of wind which racked our little barn and damaged our corn fodder quite a bit, making us a great deal of extra labor. Warren was at school, and I had much of my fall plowing yet to do and, naturally, felt thankful to see warm weather smile out again. We had a touch of Indian Summer October 31st after a very severe week of cold.

November was still very mild and still very dry. The surface of the ground was wet enough, but the springs were very low, and Willow Creek was still dry below Waukokee. Our threshing was done and we had fifty-three bushels of wheat all told, which would not amount to more than twenty bushels of clean wheat for me. The barley was merely nothing, two and one-half sacks of ground feed for me. Sixty-five bushels of potatoes and 160 bushels of ears of corn were my allowance.

The weather was warm and damp in December with very little snow. I was trying to fix up the house and barn for cold weather. Sadie Cummings, Frank's girl, came down from Worthington. Christmas brought us eighteen inches of snow, just as I was building a log shelter for the stock.

— o —

Carbuncle

The new year, 1888, found us with two feet of snow and wind enough to pile it up beautifully. Hundreds of lives were lost during January in Dakota from blizzards, never equaled. In late February a western blizzard struck us and struck hard. We stayed indoors for several days, except for caring for the animals and getting wood.

Weather was still an interesting subject in early March. The surface of our earth and every tree was covered with a sheet of ice, which came two days before in the shape of a very disagreeable rain. When night settled down on the earth, the cracking limbs in the grove sounded as though the day of doom had come to the timber, but when morning dawned, hardly any damage was visible. The vast quantity of snow began to melt by the middle of the month, slowly to be sure, but it moved and the creek was free. Stock would probably come through all right. I was trying an experiment, as I had little grain remaining. I was feeding our sheep upon the willow hedge that stood on the line to the north of this farm, cutting down the trees and hauling them to the yard. The sheep trimmed off all the twigs and then peeled the large branches of every bit of bark. Just how much nourishment they got remained to be seen.

The telegraph brought the sad tidings that Abner Colburn was killed in a railway accident in March. He was an early acquaintance of mine and a friend to all who knew him. He had moved to Oregon in 1868 or so, since which time I knew but little of him except that he was employed by a railroad company. He was promoted from time to time until he was conductor of a passenger train for several years, which duty he was discharging at the time of the accident.

April found all our snow here melted and packed down so that it would hold up a team. The farmers were clubbing together for mutual protection all over the west. I joined a society called the Buffalo Grove Alliance, which was one of many of these "Farmer's Alliances."

Mrs. C. and I had been plastering, and so everything else was put off for a while. Then came fencing, farming and beekeeping, and a lame knee all at once. My seeding was not done by May — it was wet, extremely so. One field of oats had not been dragged over, which would be a bad job if I didn't get at it soon. Mrs. C. was busy papering up. Cattle began to pick a little nourishment in the field, and it was just as well, for the supply was about exhausted in the barn. My bees were short of store too, but they appeared to be strong and vigorous.

Warren hadn't got a school yet, and we worried he probably wouldn't. When he went seeking a job in Fountain, I went to lead home the horse he rode. The wind from the north blew keenly, making it very disagreeable for me going fourteen miles out and back. He finally got a two-month job teaching in District No. 57 for $16 per month and board.

Well, summer was passing, and haying was done. Five pigs landed in my pen one morning. Lewis had to watch cattle some then. Warren helped me mow one and one-half loads of hay in the pasture. The corn began to shoot out tassels, but no silks yet.

***

Word reached us mid-summer that General Phil Sheridan lay at the point of death in Washington, and might be gone already. Grover Cleveland was again nominated for president. The Prohibition candidate was one Clinton B. Fisk of New Jersey. The Republican convention was at work in Chicago and nominated Benjamin Harrison and Levi P. Morton to run the presidential race. 'Twas said there were four parties who would cut a figure in the contest, and several side shows.

General Sheridan died August 5th in Nonquitt, Massachusetts. Thus passed from the stage of action one whom the nation truly mourned. Sheridan was a name that gave confidence and a feeling of security to us in war and peace. But after all, he, like all of us, was human.

Politics, thy name was pandemonium that year. Election returns gave everything to the GOP. They would have it all their own way for a season now. If there was any sincerity in their pretensions of virtue and sobriety we expected to see them show it. For myself, I could say I voted straight Prohibition.

It was a very stirring time that wrought political change in our country, which may be for the best, perhaps. I, myself, took a decided stand for prohibition, although we were hooted at, and called hard names, and voted down, we pledged to keep the agitation before the country until the cause was victorious, whoever held the rein of government. We had a voting strength of 300,000 by then, and the consciousness that we were right. We hoped much from our boys, as our system of education was such that every scholar's mind was bent toward the prohibition of the "great curse" that was spreading ruin over the land and threatened to undermine the government itself.

***

I hired my grain cut and I remember yet the embarrassing situation it got me into. The man wanted me to work back, which I agreed to, but he demanded the work before I could get my grain shocked, and called me a "lazy liar" because I would not go to help him. I finally got over and helped Ogg, $8 worth, and I still owed him $3.55 for cutting.

The wheat was nearly half chess. This pest kept growing worse every year until we gave up trying winter grain, but I never knew for a certainty what chess was. It grew in both rye and wheat, and resembled the grain it grew in to the extent that rye chess was larger than wheat chess, and this fact led to the theory that it was bastard wheat or rye as the case might be.

I finished my winter wheat by mid-September. No word from Warren yet. I went to an auction at John Carnegie's and helped three men thresh: M. Ogg (two days), James Bowden (one day), and G. Colburn (two days). This paid Mr. Ogg and left 95¢ in my favor. We proposed to patronize the fair that time, and I took some corn, potatoes, and garden truck. Mrs. C. invested $1.75 in a bird cage for a young singer, and our people took several premiums at the county fair.

My threshing was done in early October. I finished the potato harvest and had 105 bushels in the cellar. Warren came home from teaching having earned $23.

Sister Mary wrote us that our much-loved father Howe died, he was kicked by a colt. I should like to give a page to his memory, but not now. I will merely say I loved the quiet, kindly old man and wrote a few verses to express my love and veneration for men of his stamp. I felt like endorsing the sentiments of the reverend when he said that silence seemed the most becoming to mortals while in the presence of such manifestations of God's providence.

My fall work was hindered by an extra job of ditching and laying a water pipe from the spring. We dug a trench from the spring down to the house, ready to lay an iron pipe to bring the water. We got half the pipe in and waited a week for the balance. The boys helped some, to be sure, but it was a long, tiresome job in all.

I was shut up in the town hall on town meeting day among a lot of old smokers, and writing them up in rhyme was the only way I had of retaliating. 'Twas every word true of them.

That year marked my first real deep anxiety in regard to my final and ultimate title to the old homestead. This feeling of unrest was caused by Uncle Fowler's failing strength and his very secretive habit of distrust about very trivial matters. For instance, he trusted Jim Bowden to pay his taxes rather than me, which excited in me a feeling of doubt as to his final intentions, and I began to keep accounts with more than usual care. He was taken with a bad sore on his neck and jaw in August, and a cancer was feared. We corresponded with one Dr. F. A. Fletcher in Eau Claire, Wisconsin, who agreed that some of the symptoms were like a boil or carbuncle. He warned us that if darting pains or a creeping sensation continued, these may be signs of cancer. Of course, he could not diagnose without an examination and, if cancer treatment was required, his charges began at $50. But the doctors in Preston allayed Uncle Fowler's fears by pronouncing it a carbuncle, which was a great relief to me in more ways than one.

Warren was in Preston going to school while we at Waukokee prepared to celebrate Christmas though whooping cough threatened to throw a wet blanket over everyone's holiday.

— o —

West Point

This marks a pivotal year in Warren's life. A patriotic and idealistic young man, he sets his sights on entering West Point. This experience forms many of the beliefs that influence his political activities for years to come.

The year began with a total eclipse of the sun on New Year's Day, 1889, during which Lewis drew several pencil sketches of the phenomenon. The local alliance was electrified by a lecture from Mr. Sprague, after which money was raised to send a delegate to St. Paul to a convention of the state alliance. The Norton brothers were putting up a portable sawmill at the old site at Waukokee. Portable saw mills and steam threshing machines began to be numerous and fashionable, displacing horse power threshers and solving the problem of how to convert home-grown timber into lumber.

Sadie wrote from Nobles County that a horse had carried her safely home through a blinding blizzard. The "dumb" animal knew the way back. She was alone in a cutter February 5th when the storm hit.

A creamery scheme was on foot by the farmers, and eventually we joined the Farmer's Creamery with two cows, such as they were. This turned out to be another abortive attempt at independent dairying. I never had much faith in the enterprise and, being a small patron, had nothing whatever to do in shaping the policy of the company. They built a creamery and ran the business for two years and then failed. I was neither gainer or loser, as I remember.

Our snow was going fast. The bluebird came March 10th. Bees would spring kill badly, I feared. I put mine out one day, and back again the next. Well, what next? Farmers at work in fields, grass grew, bees hummed and turkeys layed. Winter grain was growing

171

nicely, but we had no rain—this was another feature of that unprecedented season.

April brought a dead calf at the barn, it was stillborn. My oats were ready to sprout. Warren commenced teaching the 15th. One of the turkeys was having a hard time setting. Myrtle was engaged in learning the ways of the district school again, though the whooping cough had not yet left her altogether. Miss Lela Taylor taught our school.

***

Dakota had a terrible visitation of fire, a consequence of drought and high winds. Pen cannot describe the horrors of a western prairie fire. The damage done to property was estimated at $2 million. A terrible storm at sea sent two of our best warships to the bottom of the South Pacific at a point known as the Friendly Islands in mid-March. A number of German vessels met the same fate. Strange to say, they went there to fight, if need be, and when about to part on friendly terms, were struck and rent, tossed and sunk like pasteboard toys. This was called a terrible' disaster in those disastrous times. It seemed almost like a special design of providence. First, Germany insulted our flag by firing on a merchant man trading with the natives of Samoan Island, and three of our best warships went to enforce respect to our right and met four or five German ships. Now, all but one were in the bottom of the ocean with 150 men. The one vessel saved was an American ship.

Another frightful disaster, similar to one which fell upon the people of Massachusetts fifteen years ago, sent death and devastation to the state of Pennsylvania. The report had it that 10,000 people were killed by a reservoir breaking 300 feet up on the mountains above the city of Johnstown. The reservoir was said to be the largest thing of the kind in the United States. The water rushed down the mountain into this city before the people could escape, and one of the most heartrending catastrophes ever known in America was the result. Later particulars put the number of dead closer to 12,000. Decent Christian burial was out of the question and the work of removing the dead from the debris was terribly trying to even the most firm and hardened nature. Dynamite and the like was used to blow up the jam of stuff in order to get at the decomposed remains of humanity.

***

I was in for road fixer at that time and put in a new bridge over to Facey's one day. I was one among nine men of the town chosen for the thankless task of path master at the annual town meeting. Some of the work planned by me still remains, for I changed some old "beaten paths" to better places, but I never was quite satisfied with myself as an overseer.

Many people came to celebrate the Fourth and visit Mrs. Cummings. It was very warm. Mr. Pierce and family were present. Myrtle and Lily found a rattlesnake. While they were gathering gooseberries, a big, old-fashioned, four-foot rattlesnake rose up and disputed the title to the bush over which they were bending, and under which he was coiling and rattling. The screams of the little girls called assistance, and the snake was killed.

Bees were coining in honey those days. Facey ground out fifty gallons more honey for me in July, besides his one-sixth share. I helped the Rexford boys get a bee tree hived one day. We got eight more gallons of honey, and I guessed we had better stop. I was trying to get a few pounds of comb honey and with that I would be satisfied.

Warren came home from school one day and reported finding a bee tree while on a frolic. It appears that the school went on a half-holiday picnic and, in their racing around, found a swarm of bees just taking possession of a tree on the bluff above the Root River. I went and took in that nest of bees two days later. At first, they refused the new quarters provided for them and adjourned to a plum bush where they hung for twenty-four hours, when I found them and coaxed them back with some brood from another hive. They did nicely.

Warren started school for four weeks during August. He was much interested in the West Point Military Academy and gained his first great victory at Rochester, where he won the competitive examination for cadetship over sixteen others from several counties comprising the First Congressional District of Minnesota.

Our old, first stove was discarded that fall, and $30 paid for a new one. Also the little old clock that kept time for us in early life was cast away for one of more modern type and steadier habit. Lewis took up the old one to practice on, like a medical student in a dissecting chamber where the end justifies the means. It was thrown away as useless, and laid outdoors for a season, but he got it a-running again by paying $1 to get it repaired and cleaned.

The Fillmore County fair had a balloon side show that year. We saw it from here as it sailed off south and went down.

My stepmother "number two," David's third wife, died in September and was carried by the drunkards to the Prairie Queen cemetery, where they preached her funeral. Father came to live with us. Warren came home again to help nurse a poor, sick friend. He put forty-eight bushels of potatoes in our cellar as the result of his first attempt at independent farming.

On Reuben's birthday, a thin snow lay upon the ground. Our threshing was done with steam power. Warren and Uncle Fowler ran a water spout to the barn. The boy was teaching in District No. 131, and Waukokee took two weeks vacation.

I may as well make a note of the fact in big capital letters that Doc Jones took out that old wisdom tooth for me using a monkey wrench, or an alligator wrench of diminutive proportions, to perform the operation. I was not permitted to scrutinize the implement very closely, but must conclude this by the sensation I experienced in the operation. At last, this time he succeeded in extracting the wisdom tooth. It had tortured me until I just had to try the cold steel on it again.

One item of great interest for a short time was a runaway. The horses and I started from above the spring with a load of wood. All went well until an unlucky slue or lurch of the sled sent me and the load into the ditch and the horses and sled into the barnyard on a double-quick, where I expected a general smash-up, but found all serene and right-side up except a broken strap or two. The incident caused me to buy new lines and ever after to keep a sharp lookout for weak spots in my harness.

A new disease was among us which was very troublesome and quite dangerous. I contended until then that it was more talk than anything else, but when death called for a man we must conclude that he was dead in earnest. It was a foreign disease called *La Grippe*, or influenza.

Warren finished one job, teaching, in late January and began another in February, posting up for examination. Warren came home and he looked sorry; if he didn't mend fast, he wouldn't get to his examination.

We had two sick cows that I gave up to die, but they were right again after a week. They had filled up with bran, so I gave them four quarts of hot water and salt as strong as it could be made. I had found this a remedy before, but the cows were sick thirty-six hours before I knew it that time, so I didn't think it would work. Beef was very cheap. I butchered two that winter and was trying to eat as much of it as I could. Since 2½¢ per pound was all I could get for the hide, I was trying to tan it for a lap robe. It was black, and if I could make it soft and nice, it would be worth $5. I sold one quarter of the beef to Warren for his saddle, and he got $5.25 for it.

During February, quite a sensation was caused by what would go down on history as the "Women's Crusade." In Missouri, some exasperated females made a raid on a number of grog shops and left them completely wrecked—Carrie Nation came into national notoriety as a saloon smasher.

We had an institute in Preston that I attended one afternoon. John St. John lectured us on Prohibition. He chased the tariff humbug into its hole and plugged the hole most effectually. I wrote a temperance poem entitled "Watton's Folly," which would have passed unnoticed as a bit of pleasantry but for the fact that the death prophesied for him fell upon his wife in a case of heart failure, which had been predicted for him. This turned my fun into seriousness, but truth is eternal.

In late March the schoolteachers had an institute, which Warren took in, and Miss Myrtle attended a part of the institute, too. Rose had a visitor who jerked out her teeth and such. If I remember correctly, she had four less teeth and three of those remaining were filled, besides. A carpet was under process of construction, and we were on our way to glory.

I saw some dehorning done and was of opinion I could do the job. I tried it later in the year. There was no one to tell the tale, for I did the job alone with the aid of a forty-foot rope and three ten-foot ones. For "further particulars call at our office." The stock were doing nicely, and I was very well-satisfied. I was reading Josephus that spring. Rose was sick, "grippe" again, I guess. I sold a hog and gave $10 to Warren for his work on the place.

Our family picture was taken in the spring of that year, before a break was made in the "circle," as Warren was about to go to try his fortune applying for West Point. When the portrait came back, "Good enough, all but me" was the general complaint.

Warren went to the West Point examination in New York in June. I gave him $16 and signed a $30 note with him, due to L. M. Conkey. He left me his saddle, watch, and organ to hold against his safe return. He was home by the middle of the month. The miserable enterprise was finished and a failure, so far as his cadet-ship was concerned. The venture proved an empty honor, with much expense, and ended in disappointment at the last. Political preference seemed to rule the patronage of that institution. But, I saw no cause for regret at the attempt. We dropped the curtain over West Point, said it was all well and good, and we were satis-fied to drop the subject.

Several years later, Warren wrote a letter to the editor[*] about the experience, to warn other boys about getting their hopes up and dashed:

Those West Point Examinations

The Twin City papers mention competitive examinations for West Point to be held soon under the auspices of two distin-guished congressmen to fill vacancies at the military academy from their respective districts. Here, I feel it my duty to say a word in this connection for "none can speak as him who from experience talks."

In 1889 I won an appointment to West Point over seventeen competitors. Attempts were made to bribe me, to scare me, and to discourage me away from the project, but I stood by my guns and reported at West Point for preliminary examination in June, 1890, pursuant to instructions from the War Department.

---

[*]   *National Republican*, April 12, 1894.

Physical examination came first. Clothed in a firm resolve I sailed through this; muscles, bones, joints, vital organs, eyes, ears, and all with no difficulty until the medical board reached the base of my nose. Here the learned gentlemen called a halt, held a brief council of war in a subdued tongue and then wrote a report, sealed it, sent me out of their august presence and into the presence of one Adjutant W. C. Brown. Mr. Brown broke the great seal and read to me my sentence which was: rejected on charge of "imperfect physical development and nasal polypus." Now, dear reader, how many people are physically perfect? The great majority of boys who were examined with me that day were accepted, as I understood it, yet they were not modern Apollos by a considerable. As for the polypus, several reliable physicians prospected for it shortly afterward and reported no find.

Of course, the whole business was a fraud. It cost me dearly, but it was worth it to me as I learned to view politicians in their true light. It taught me that our country is in greater danger from political corruption than all other foes combined. "Educated soldiers of the ballot" is the crying need of our land and it behooves every patriotic citizen to get to his post and do his duty in preaching the truth if we would preserve the rich heritage of liberty and prosperity delivered into our hands by our forefathers.

Now boys, if you want to see the grandeur of our natural and civic proportions, go to West Point and keep your eyes open all the way down and back. Don't fail to ride down the Hudson to New York City. The palatial residences which adorn either mountainous shore of the noble old river bespeak the awful story of millions living in poverty and drudgery while a few absorb all the wealth their labor produces.

Would you know how the few reap the harvest grown by the toil of millions? Let me whisper to you the secret. Two things are essential to the prosperity and progress of civilized people: the first is production, the second is exchange. The millions produce, the few control exchange of products by controlling the two great instruments of exchange, viz., money and transportation. They control money and transportation by controlling politicians who are elected to legislative bodies and appointed to our courts. They control politicians by giving them a very minute share in the spoils of this terrible war against mankind.

—W. E. Cummings

The alliance succeeded in breaking the twine trust—we got twine for 10¢ that year, which had been 17¢ the previous year. I paid a $5 bill for fifty pounds of twine.

I heard of so many cyclones that I quit making a note of them, but one I must speak of: it struck an excursion boat on Lake Peppin, and over one hundred pleasure-seekers went down in death. There were other places visited, of which St. Paul was one, and it was reported that Stillwater was struck by one in the middle of July.

We were in the midst of collecting basswood honey, the first honey of the season. Quite a raspberry harvest was being gathered by our people also. It was very hot and dry. The ground was baked hard, and oats were rusting badly. Warren went to Mower County to do harvest work there. Our harvest was gathered in by mid-August, and then we helped three neighbors thresh.

Father stayed with us until August, and I suppose the less said about his eccentricities the better. He was seventy-seven years old and scandalized the family by starting a stubborn hunt over three townships for some old woman that would marry him. He played the fool all spring and summer and we thought he had quit. After remonstrating in vain with him, I gave him over to his folly, and he married wife number four, after which I resolved to have no more to do with either him or her. This may not be justifiable, but it seemed best to me and I could not find it in my heart to condemn the course.

Warren went to Preston in late August to begin study. Lewis put in a few days at Dayton's, then helped Long thresh, then went to work for O'Brien. We had a Prohibition convention, which I attended, but gave up the enterprise after a few hours to come home to my chores. We went to the fair in September and saw balloon ascension. I, with the rest of the crowd, saw a man go up into the clouds. I had no desire to see it repeated, nor any other fool play.

The alliance started getting involved in politics and put a ticket into the field independent of all other parties. Although their platform savored slightly of prohibition, it was not enough to warrant a prohibitionist supporting the party.

Warren came home. Politics seemed to be the all-absorbing topic with him then. He was a regional delegate to the state convention that year. I wished he looked out more for his own interest, but all our county was being stirred up by political agitators and he was being used by them to help do the stirring, which I feared would result disastrously to his reputation. But, my advice in the matter seemed lost in desert air, so I resolved I must do as I did with the gun: warn him that it is loaded, tell him it will kick, and trust to luck for his safety.

When the election was over there was quite a revolution. I learned that Warren got his vote challenged and was not surprised, as he took a very active part in the squabble that ended in the election of Captain Harris over the boodle candidate Dunnell. Warren had good reason to suppose that Dunnell had sold him out the previous summer at West Point. This election acted as a sort of balm to the boy's wounded feelings, but could never repay him the sad disappointment and loss of time and money that this unscrupulous politician caused him, just to hoist himself into position. Now Mr. Dunell could go home and nurse his own disappointment and know that his treachery had its own reward.

I was alone again for the winter. Lewis went to town to begin study in the high school. I took in my corn and husked as opportunity offered. Uncle Fowler and I were quite lame with rheumatism.

My record for the year was full of blanks as well as shadows, and I will make no further mention of them. One thing I will note a little, our Farmer's Alliance was about busted and I, for once, was disgusted, for on broken straws I'd trusted. They ran into political pollution and were pulling wires after the fashion of corrupt party bosses. I had seen my hopes crushed on every hand so much so that I merely worked on mechanically — like a machine — because I had to, but expected nothing to come of it but weariness and vexation of the spirit.

The day after Christmas was my butchering day. My old cow was converted into beef for home use; it didn't pay to sell beef that year and I only got $1.82 for the hide. The holidays brought us their usual allowance of presents and pies, with the bellyache accompaniment.

— o —

Dark and Dreadful Cloud

Fred writes very little about Rose in his journal, but it is evident from the little he writes and his very early and very late poems that he cares for her deeply. They are about to face a crisis that will take Fred and Rose further from their home than they have ever been before.

New Year's Day, 1891, gave us a blizzard. The boys were both home, and we read about a horrible Indian war on our frontier that had been in progress for six weeks. I began to be glad that my boy was at home, instead of at West Point. A "new thing under the sun" was a "phonograph," which I will not attempt to describe but will just note that the boys were experimenting with the thing, which was being exhibited in town as one of the wonders of the age.

Steady cold weather and *La Grippe* prevailed. I was engaged getting up summer wood and putting in spare moments rewriting "Rural Gems" for the last time, after which all mistakes are to be considered such to the end of time, if the ink holds out so long. This was a collection of many poems from over the years, and I called the bunch "Rural Gems" or "Crazy Patchwork from a Plowboy's Pen."

The boys were at the teacher's institute for a week and then attended the examination. We had a good pile of wood up at both houses and a big pile of manure in the yard to haul out next. Easter brought the robin and bluebird on March 29th.

Lewis began a job at 75¢ a day, working on Facey's farm. I contracted with Facey to tend my bees again. Myrtie took music lessons from Mrs. Phillips in Preston. I was trying to read Byron that spring. He soars too high for ordinary minds to grasp his meaning in all its fullness, but I managed to figure out some things of considerable interest from his illustrations.

I was almost in despair over matters and things of no particular importance to others, but they weighed me down entirely beyond my endurance. I believe, were it possible to run away from the torment I should not have stayed here long enough but to load my wagon and leave. Instead, I lost myself in books and whiled away a beautiful spring reading Longfellow.

Sister Vina returned home with Myron, who appeared to be a fine lad. While here, he engaged in trapping rats. Shortly after they left, Vina wrote that a sad accident befell Myron, which caused the loss of an eye. By then I was reading *Talmage in Palestine*. Dogs were a terror to us that year due to hydrophobia, also known as rabies. The mad dog scare in the vicinity made me shoot Dan Thatcher's dog and a miserable, mad scrape followed. I gave him a shepherd pup afterward, which partly compensated for the crime.

We went to Preston the Fourth of July, and such a celebration! More company came to visit: Grandma, William and Christina, besides Lily. Warren finished his teaching job and took another immediately. He was up to his ears in papers and business in the printing office, too.

I was reading a partial history of "Great Caesar" then—Great Butcher, if you ask me. I read A. L. Sleyster's poems, too. A skunk visited us two nights in a row. I was almost laid up with a sore side, neuralgia, but had to keep moving.

A big shower came the first of August. I sold a calf for $3.50 and, being caught in the wet, spent it all for coat and umbrella. At first glance, it would seem to have been a bad trade job, but an umbrella was always in order for Myrtle, and the coat was better than a siege of rheumatism. My lame side still made me much trouble. Rose had the pride of owning the nicest White sewing machine in Waukokee.

Haying was at last ended, and my wheat crop was gathered, five loads. The grain was good, but oh, so much chess! There followed another week of hard work and perplexing circumstances. I commenced plowing Monday, helped John Dahl Tuesday afternoon, helped A. Long Thursday evening, and tried to stack Friday. Saturday, Lewis moved to town. With these brief interruptions, I had also been turning soil. We had several frosts, with no apparent effect on crops, and a rain at night with the usual depressing influence on already damp oats and young pigs, of which we had eight. The outlook was not very promising of ease or comfort to me, but still I held on my course with a tenacity born of desperation. I knew I was in for it and would fight it to the bitter end.

Myrtle and her mother were preparing for an exhibit at the fair in Preston. Mrs. Cummings went to the fair and took twenty pounds of butter. Myrtie got a premium or two. Warren went to the state fair in St. Paul as representative of the *National Republican* newspaper of Preston. He came home again for a respite of a few weeks before his school began. His brief career as editor of the *National* seemed to have given general satisfaction. Lewis got his teaching certificate and was negotiating for a school.

The balance of my time was devoted to the potato harvest mostly, while Warren was breaking up the east ravine and threshing. Facey said our bees had to be fed, and I gave nine gallons of honey to Facey for my half of the fodder. This was rather steep speculation — money out of pocket — so I pledged to tend my own bees next year.

Warren began teaching in November, and he bought a little farm — thirty-five acres. Lewis was hired to teach a four-month school for $30 per month — 'twas his first attempt.

We were notified by mail that a young lady was to be at Isinours the next day and we were expected to meet her, greet her, and bring her home, no more to roam. Her name was Maude VanSickle, from California, whither she had been carried by sister Kate. We cared for her about a year. A Vermont cousin, Henry Cummings, visited us in November — just one day, and went home. He lived only about ten months longer, I believe, so I was thankful for that one day.

I was still "down in the dumps." My work worried me almost to the point of insanity and the weather didn't help me any. I was ashamed to admit how my water spout froze one night. To cap the climax, I dreamed of muddy water. Monday was Alliance Day in Preston — I took it in — and a woman spoke. Lewis kept me in reading matter then, as he had access to a library: *The Saracens* and *The Greek*, with a novel, *Ben Hur*, so far.

Probably I had never before sat down to write with so many conflicting thoughts and emotions running through my mind. If my feelings were consulted, I would far rather be wandering out over the bare, frozen ground, letting my thoughts run wild, than to try to compose myself enough to write one sentence that would express my feelings on the first Sunday of the new year of 1892. And I decided wouldn't attempt it.

The whole neighborhood had the grippe, or expected it. A dark cloud was rising over our family horizon, but we hoped it would vanish before it reached us—something that Rose had noticed a month before. I was reading the history of ancient Rome and Greece, and then found Byron's *Childe Harrold* very interesting.

The middle of January was thirty-six below, but after that it grew warmer, which threatened to despoil us of our little snow. I was trying to save some ice for summer. As this was a new enterprise, wasn't sure how it would work. I cut big blocks of ice from the creek, put them in a dug-out ice cellar, and tried to use straw as a preserver.

I went to town in early February, with grist and twenty-eight pounds of chicks, and then again with 780 pounds of steer for Uncle Fowler. I had five papers sent to me, besides *Gleanings*, and it made me feel poor. One of them came gratis, the others were $1 apiece.

The middle of the month found a calf and a lamb at the barn, three pups in the woodshed, four loads of ice in store and room for more. I was putting "Blizzard Joe" in better shape, and was half through by then. A week later, I had finished my ice bin, also my poem. It was probably my longest piece, and perhaps my best effort of any note. It was first written many long years before.

Mrs. C. and I coughed and sneezed incessantly. We had been sick for a week with no signs of mending: *La Grippe, La Grippe*. The weather was foggy most of the time but, at last, the roads were bare again. *Peter the Great* was on my mind and my reading list then.

Lewis went to try school again after a week's layoff, and Myrtie was reported sick at Isinours. There was trouble enough in my ranch, and oh! that cloud, that dark and dreadful cloud—would it ever reach me?

My bees furnished us with a $6 Bible, received in March. It was a big Bible, bought of Facey for a colony of bees and it was very nice as well as instructive. I also sold $4 worth of honey to Conkey Brothers. The weather remained mild, and *La Grippe* was gradually letting up.

The bluebird came March 7th, followed two days later by the worst storm of the winter. I was roused at three o'clock that morning by the rain, and got up to shut up the sheep and bring in some wood. While doing this, I heard a roar in the west, and by four o'clock the rain had ceased and snow began to fly before a blustering, cold wind that grew more so as the day went by. The mud was very deep. The storm caused much suffering far and near, and many deaths. My strength was slow in returning after sickness; and Rose, my poor Rose, God pity us was all I could say. Lewis and Myrtle were all home at the end of March, with school out for two weeks.

We finally had a name for that dark cloud. Rose was stricken with breast cancer, first noticeable in December 1891. A ray of light penetrated the darkness and we hoped still, and proposed to go to Wisconsin as soon as possible to try the skill of a specialist who advertised to cure cancer, and our faith was strong to accept the situation and take the chances.

We reached Eau Claire, Wisconsin, on April 4th, in a revival season and were greatly blessed by the meetings and the sympathy of Christian people. These memories will stay fresh in my mind when the care and homesickness that accompanied the long, anxious waiting are forgotten.

Dr. Fletcher, a specialist in cancers, agreed to give treatment. We found the family very kind and considerate. Although we were homesick, we settled down, accepted the inevitable, and tried to get acquainted with conditions and fellow sufferers. I saw many strange things and strange faces of all shades, and four races: Chinese, Indian, Negro, and white man all mixed up and associated together. Hundreds of people thronged in the streets. Many saloons kept open all hours of the day and night, and some didn't even observe the Sabbath. We stayed in rooms so near the tracks that a shrieking train whistle woke me every dawn.

The patients were given morphine pills, one-eighth grain, at the doctor's discretion, and I will say there was danger of a habit forming at that dose, but Rose mastered it by stopping as soon as possible. But its after-effects were a source of anxiety to me. The cancer came out on April 21st after it was killed by plasters and drawn out by flax seed poultices. The process and treatment kept us there until May 1st, just a month from start to finish. Then we went over to Augusta, Wisconsin, to visit relatives for several days and back again to get ready for home at last. I paid the doctor $50 — all I had — and he would take no more. The trip cost me $75 in round numbers. I got off cheap to what some of my fellow sufferers did, but it was all I had and Lewis furnished most of it, with Warren contributing some too. So we began again at the foot of the ladder, thankful if the trouble would not return.

The last of May was a hard time for farmers. There was snow in the morning. We planted early potatoes and tried to plow. I went to town, got some stuff for salve, and made the mixture for Rose. Lewis was all busy with a fence machine he was inventing. The apple trees promised much and the bees took their share. Rose was gaining strength day by day and was almost well.

Another stormy week; I finished planting corn in the rain a good deal with the hoe on account of mud. Lewis managed to be away all week one place or another and planned to be away half of another week. I planted some potatoes but had most of it yet to do, and planted pumpkins in a patch by themselves.

Warren was home looking after his property. Aunt Net came for a visit, and Miss Maude was here. Net told us that B. F. Cummings was lost to his family, and had been for two years, but I didn't know it till now, and it caused me many a sad hour. I took long walks before breakfast. I planted a big bean patch, sheared sheep for Miner, and planted some potatoes. Lewis completed his fence machine and thought it was about perfection.

I was more or less sick and didn't know what was the matter — guess the wind was in the wrong quarter. It was raining again steady and straight, which was a common sight that year. Fair weather was a rare occurrence. What will the cornfield be was the question, soon solved, for it could not stand much more wet.

Jennie Taylor was here, and Warren's school would be out in one week more. Rain, rain, steady and straight all day with no letup.

***

Politicians were busy and four men were pitted against each other for presidential honors. Ben, the Republican, and Grover, the Democrat, ran for the old parties; while the Prohibs and the People's Party tried to keep their end up with Weaver and Bidwell as leaders. I was just swept along by the great onrushing tide, protesting and helpless. Weaver was as much a Prohib as Bidwell, and the Prohib platform was a more sensible and businesslike production than was the People's. So what could a fellow do but cast a vote for God and prohibition and let men rage and imagine vain things? But, I feared a bloody revolution if things continued in that state long.

There actually was a bloody riot in the steel and iron works at Homestead, Pennsylvania, with the killed and wounded reaching up into the hundreds.

***

Warren went to get his things. He took the Harmony school job for eight months at $55 per month to begin in October, and was trying to start a paper in the meantime in the interest of the alliance. The harvest was on, and Lewis was working for $1.25 per day for a man who would also cut my grain. Warren helped me stack the loose hay, and Jennie was here for a visit again.

The following week I cut and put up the remainder of my wheat after binding about half of it. The stuff was dry and we proposed to stack it the next day, if all was well. Warren was helping with a saddened heart and a maddening sensation in this brain, and we all sympathized with his feelings more or less. I went to town and my heart ached until my head whirled. It seems as if I carried that burden of suspense and vexation as long as I could, and I felt anxious to find a spot to rest just a brief while at least. This is incoherent language to those unacquainted, but to me it is all too plain.

The anxiety ended very happily to us all in the union of our Warren and his Jennie. They were married September 27, 1892, and seemed very happy and contented with each other. October came again with its golden leaves. Warren and his bonnie bride left for Harmony—he to take charge of the school, and she to keep his heart.

Lewis and Myrtie went to Preston to keep house and go to school. They rented three nice rooms for $3 per month. This made two Sundays that I was on the road and I didn't like to spend my Sabbaths in that way. The young folks from Harmony came for a visit, and it was clear that Warren and Jennie were happy. The children at Preston were doing well also, except that Lewis came home for a few days, quite sick. I returned from Harmony after taking Warren and Jennie home, tired of trying to get rid of the burden imposed upon me. I took it up meekly, and said nothing. Sister Kate was there, and her three girls here.

A very cold, windy week passed, also three large flocks of geese. November 7th was election day and we voted, according to law, by the new system. Grover Cleveland was made president, but I voted straight Prohibition. Lewis was twenty years old, and we celebrated with all the VanSickles here, also Warren and Jennie.

Rose had been worrying about her cancer again for a couple of months. In late November she and Myrtie went to see Dr. Fletcher in Wisconsin again. I heard nothing from Rose and was very much disappointed in not getting a letter. This brought a peak to my sorrow and anxiety, but it proved a false alarm. On December 11th I received a letter saying she would be home in three days. A soothing application stopped the irritation and allayed her fears. This cost me $25 more, and Uncle Fowler donated $5, for which I was grateful.

Mrs. C. and Myrtie came home December 14th, much to my relief, with the doctor's assurance that we would not be troubled by that cancer again. Uncle Fowler was sick and his chores were added to my load. The weather was terrible, I thought as I went to Harmony for the children. Warren and Jennie came home for two weeks at Christmas, and Mrs. Miner and Minnie came by for a visit. Lewis took care of Uncle Fowler, and I went to get the doctor for him.

Religious experiences of 1892 were perhaps the most marked of any year up to that time in my life. My family afflictions had much to do with my state of mind, no doubt. In times of hardship we realize our dependence on God and, in doing so, become aware of the fact that it is necessary to get right with God. Affliction seems to be God's opportunity to console and heal as well as to convince and convert.

Much has been said and written about answers to prayer and some are apparently trivial and foolish in proportion, no doubt, to the amount of mental capacity possessed by the writer. Be this as it may, I gladly confess that I was at my "wit's end" at this time, and the Lord heard my cry and helped me out of all my affliction. Not only that, but I hope it thoroughly cured me of being an infidel, with all its various forms and phases, for they are legion with their plausible arguments, which at best only show the ingenuity of men.

— o —

Aunt Lydia

Six visitors came to our house to celebrate the new year, 1893, which opened on us quite pleasantly though it found us sick and well about half and half. I was very sick, I guess it was grippe. Three visitors lingered well past the middle of the month. Warren went back to work and Myrtie went with him to work for her board and go to school, which made me very sad and lonely. Lewis left to teach school and board with Mr. Facey near Fountain. I was kept in close confinement at hard labor, trying to keep the work done at the barn. Uncle Fowler appeared to be failing, and we were greatly relieved when he started improving after several weeks.

February brought a terrible week of snow and blow — the snow drifted until it couldn't drift any more. Lewis came home, announcing that he had a two-month school at $35 per month away in Chatfield. They were all sick at Uncle Fowler's and I had more than my hands full doing his chores for another week. Mrs. Colburn, Uncle Fowler's sister, was very sick. Aunt Lydia said that she was not sick; "But," she said, "I am far from well." Uncle Fowler was some better. Charles VanSickle visited here and the sick were better for a few days, then Uncle Fowler had a relapse. Aunt Ruth Colburn left with Mrs. Freemire.

I found I could run my cross-cut saw alone and did a very good business, yes, and worked just as hard as I pleased. I finished the old maple and was glad to see the last cut come off, but it was a sleek job. There were about eighteen cuts from twelve to eighteen inches across.

Uncle Fowler got better, but Aunt Lydia took to her bed. The February wind blew dismally and made a homesick howl around the corner of the house. The Harmony folks, Warren, Jennie, and Myrtle came over. We went to town and brought back Aunt Ruth.

191

She was better again, and all the other patients at our hospital were comfortable.

It snowed as I took the young people back home, and I never saw snow fall thicker than it came down when we were in Harmony. Then, to change the dull routine of snow and blow, it set in and rained after I got home and blew "Hail Columbia" from the northwest at night.

***

The papers said that ex-President R. B. Hayes had died, and General B. F. Butler also. Two men that did as much for us as they could to say the least, and perhaps we can safely say no two men have done more. Sturdy, honest, old Ben Butler! Mild-tempered, kind-hearted R. B. Hayes — a man who was brave enough to banish king alcohol from the White House. We must think of them as gone forever, and in a home beyond the river. Ex-Secretary James G. Blaine died in January, and he was too well known to need any of my "eulogium." But I will say that I might have liked him better if he had taken a more manly stand against the liquor traffic.

Grover Cleveland was our president then, inaugurated in March 1893. The country was in the midst of a money panic and a revolution was threatened in the financial system. The Eastern money kings were trying to have silver put on an equal with gold, as it used to be. Free coinage, said the West. Demonetize, said the East. Robber, said the West. Crank, said the East. We shall see what we shall see. This big fight over the silver dollar kept on through October, until it was over and lost in November.

***

Myrtie went to Preston that spring, going to school, and Jennie was home with her ma. Warren was back in Harmony, Lewis was at Jim Bowden's, and I at home setting turkey eggs. The snow was from six to sixteen inches deep and still coming down. After four days of storm, snowdrifts from three to five feet deep covered the ground and it was eighteen above zero on the morning of April 23rd. One of the horses, Dan, in a playful mood gave me a severe kick in the chest. The snow lingered and it stormed some every day for four days, so there was no weather for farming that week. I was still confined to the house and expected to be for many days unless there was some change, both in the weather and the sore chest I had. It tortured me most unbearably at times. I was unable to labor much and slept only by being propped up in bed. Aunt Net was visiting again with three; Sadie had two children now, one boy and one girl.

In May I learned that W. N. Tinkelpaugh was dead in Montana—he had been our neighbor twenty years before, then he went west. I finished my oat seeding by the middle of the month, which was probably the latest seeding I ever did. Twenty-one little turkeys hatched off those eggs I set, and we set many brown leghorn hen eggs for Warren, who was a pa since May 12th at five minutes past eleven at night. Our Jennie was an angel to take care of, and her babe was like its mother—she was named Myrna Violet Cummings.

About this time we surmised Aunt Lydia was dying. Sister Kate came, hoping to be in time to be recognized, but I don't think Aunt Lydia opened her eyes after she came. I don't know that she could. She said, "I suppose I could if it was necessary," but when Mrs. Colburn invited her to, she said, "I don't have to." Aunt Lydia died May 30th. Brother Bailey preached the funeral from Colossians.

A few lines given to my aunt's life would hardly be out of place. Born August 26, 1818, she knew the "pinch of poverty," although she never went hungry or cold beyond endurance. She learned the value of rigid economy. With not much chance for schooling, she became a fairly intelligent reader, her accomplishments rather above the average. The other common branches I know but little about; except grammar and geography, which she seemed to think entirely unnecessary for common people. She always worked hard, until stricken with paralysis in the spring of 1878. She had a remarkable, sweet voice, and loved to sing when alone. Of her religious belief she never said much, but was a constant Bible reader. And of her kindness and goodness of heart, she needs no words of praise, her constant sacrifice for others proved her virtue.

But for all her kindness and conscientious, Puritan ways, Aunt Lydia was painfully set against human progress so far as either herself or we young ones were concerned. Hence, all the education

I got in the way of the higher branches, even grammar and geography, I had to fight for, figuratively speaking of course. Singing school lessons were a senseless jumble to her, and music lessons were not so much as thought of as among the possibilities. Yet I don't find it in my heart to complain or find fault, for I always felt unworthy of the many advantages that I had, and my conscience daily reminded me that I was getting more than I deserved. When she died my soul seemed lost in doubt and despondency. The only mother I ever had, Aunt Lydia had been helpless for fourteen years. My constant care and daily companion, not a day passed but I visited her if only to see that she was in her usual health.

Now, all was changed. She was gone. Where? Love said the angels took her. Doubt and despair said "never more." We knew nothing of a future except what we found through faith. Turn whichever way we would, the theory of a future state was empty speculation without the Bible. Then why do we not study it more for ourselves instead of taking it second-hand from other men as weak and fallible as we are?

— o —

A New Cornerstone

As during Waukokee's earliest days, if a church was to be built, its members pitched in to do it. This time, Fred and his sons are in the middle of the action. The building served them well for decades, until it became hard to support a country church. Preachers in those days were often paid through firewood and food given by the congregation, with occasional collections taken to provide money for clothing or a small stipend. A hundred years later, nothing is left of either the Waukokee school or church except a few foundation stones.

Our baby, Myrna, was a little gem. Rose prepared to go to Owattona to see her sister Vina, who had been terribly burned in an accident with a gasoline stove. So the end of July found me alone so far as wife was concerned, and I was very busy all week haying. The barn was nearly full, and most of Warren's hay was stacked. Warren moved his household goods back from Harmony to here. Harvest was gathered and mostly stacked around the neighborhood by mid-August. Mrs. Cummings didn't write and Mr. Cummings was quite weary of such a state of affairs. My honey harvest for the season footed up to about ten gallons, and I gave Brother Ravell some. Our baby was growing in grace and knowledge every day.

Mrs. C. came home at last, so did Lewis. I was tired and nearly sick from hard work and eating too much of apples, plums, and grapes. We had been laboring a week at digging Warren's cellar, and it was not yet done. September was hot and dry. Myrtle and Lewis took in the fair, and I attended it the last day.

Lewis bought out his grandpa for $500, and David prepared to go east. He was eighty years old, and yearned for the home of his younger days. I went to town one day to fix up Lewis's business with the bank over the loan to buy Father's farm. This transaction gave me more or less worry, and some small satisfaction. Father David started for New Hampshire and Vermont. Years later, Lewis sold the place, doubling his money and releasing me from all further obligations by paying off the indebtedness.

I sold a cow to Facey for eight weeks of Myrtle's board and worked two days on the Facey hill road. The rest of the time was spent husking and cleaning out the wheat from the oats, which I found to be a hard task indeed. Warren and wife went to hear the new preacher, Ed Thomas, and Lewis worked for him a few days.

Rose and I went and fixed up Aunt Lydia's grave on the first warm, nice Sunday in November and then went to church. I put in several days for Warren that week, and his house was finally square over its foundation.

Our cows were turned out on the cornstalks, Warren and family moved into their own house, not too far up the road, and he started another teaching job. Lewis turned twenty-one, and Rose and I had our twenty-fifth wedding anniversary. The neighbors brought us some silver, and we had a party.

Winter was here by Thanksgiving, and as usual it made us hustle. I butchered a three- or four-hundred pound hog, and two of the same for C. P., which meant a lot of pork to salt down. Uncle Fowler was lame, and I had his chores to do. A rainy, disagreeable day carried all the snow away and left us in the mud for Christmas. We had many VanSickles here visiting, which was not altogether pleasant.

In looking back, it was noted that the silver coinage battle was fought and lost in 1893. I am not ashamed of my part in the struggle for the double standard. When the fight was lost, the little squib I sent to the paper deserved better recognition than it got when compared with the worthless trash that went to make the bulk of the common county newspaper. 'Twas a little poem titled "Gold Is King," and began:

> Aye, Gold is King, let all submit
> And bow the head and bend the knee
> Like craven curs, we're only fit
> To live or die at his grim decree.

Myrtie started at Waukokee school in early January, 1894. Lewis and she were having a terrible time with a cough that winter—I don't know which coughed the worst. Warren and Jennie, ditto, but baby Myrna was well and a most lovely child. We finally sent for a bottle of medicine for Myrtie toward the end of the month. I was sort of sick by then, too, as my digestive machinery was wrong some way.

The Waukokee schoolhouse was kept warm by evening meetings in the charge of Rev. Hawke—they proposed to take in members and reorganize the Methodist Church here. Mrs. Cummings and I proposed thereafter to help maintain the Methodist Church at Waukokee.

Rose got sort of sick in February, same as I had been the previous week, with a bad cold to boot. It seemed that our people were doomed to cough all winter. A week later, Mrs. Cummings was far from well and I got two more bottles of medicine for Myrtie. Jennie was here most of a week, sick with a sore throat and a terrible cough. Little Myrna was also quite unwell by then, having a bad time with an earache.

The first bluebird was heard March 3rd, and a flock of geese flew over in the evening. My poor Myrtie was still a very sick girl, and the doctor came to say she had pneumonia. She lay very near to death's cold river, we were making a hard fight for her dear life. In our weakness and ignorance we feared we were about as apt to do the wrong as the right thing for her. Dr. Phillips came three times and we were worn and weary with watching, but would count it all as joy if she could be spared and regain her health. Our prayers seemed answered as Myrtie was so much better by the end of March that we were sure she would soon be well.

We received word that my cousin, Henry A. Cummings, had died in Thetford, Vermont. I bid Mrs. Frasier good-bye when I visited her, knowing I should probably never see her again in this world. Even Dr. Phillips was quite sick and Lewis was down with measles.

Myrtie was improving every day. Warren's folks still held out against disease, but I was sick next. My trouble started by taking cold April 7th, then the rash set in the 9th, and I tried to keep about and work it off. I succeeded in keeping on my feet, but came near losing my life for my carelessness, or obstinacy, or whatever it may be called. Still, I had to work and trust to luck to help me out. Then Warren's people came down sick with measles. I was still unable to do anything of consequence. I never had such a time with my nose—it wouldn't let me breathe a clear breath for two weeks.

I went to the mill the end of April with three bushels of stuff which was my last grist for the season. Baby Myrna was sick with measles and had it quite hard. I was plowing the hill east of the road, and it was a hard job, keeping going through all that illness. May brought Jennie's 21st birthday, and Myrna's first. We went to the cemetery and fixed up the graves of our beloved dead.

We finally had several nice, warm days. I planted my potatoes and had about six bushels in the ground. June opened warm and dry, and the corn looked well. I was very weary. I worked hard all week shearing sheep and working on the road. I dragged over my corn ground also, and helped to bury a wee baby on May 30th. Lewis went away with Jim Bowden for a tramp through Iowa. I went to another baby funeral in forenoon, Alfred Ayers's child that time.

We had quite a lot of rain in June. I plowed most of my corn with Billy and the shovel plow. We had not yet heard from Lewis. July turned hot—ninety-four in the shade, and corn was way up. Lewis returned, and announced he would work for five months at $20 per month elsewhere, and Warren was engrossed in political nonsense.

***

Over the ocean, the president of France was murdered on June 25th, and our own land was in a state of foment such as had never been seen. Strikes and tramps threatened to overrun the country and defy the laws of the land. The reason for it all lay in the fact that the rich trampled on the poor until the oppression became intolerable and in their efforts to break the yoke, the downtrodden laborers overreached the bounds of reason and were likely to make a bad matter tenfold worse. God pity America when fools and knaves hold rule, for anarchy and riot follow in oppression's wake.

A big railroad strike was ended by U. S. troops. It was hard to get the straight story of this affair, but the action of the troops was strongly condemned as military despotism. Another noted poet laid down his burden and entered the great mysterious future: Oliver Wendell Holmes, eighty-five years old.

***

July 22nd was my birthday and I was forty-eight and weary, but was trying to repair my cyclone cellar. It was ready for the rafters but I had to go off to do haying, and finish the cellar as time permitted. Harvest was here and brought quite a fair yield, though it was very dry. Pastures and potatoes cried for rain and the flies were a terror. Basswoods seemed to be yielding considerable honey. It was still dry, but much cooler, with quite a severe frost in early August. My harvest was finished by Alex Long and I stacked with Tom Yeast's help. I was very happy when Yeast was done with me and I with him.

I may as well record the fact right here that I felt the need of spectacles to write with for the first time. Threshing was done by early September, and we had 301 bushels of mixed wheat and oats. I bought James Bowden's marc colt and two-horse buggy for $45, giving him my note to that effect due in two years. Some of my potatoes were dug by October, and were good but few. They were planted too deep for convenient digging so I had to plow beside the row and then pull out the crop with a fork.

October was very lowery with not much rain. One of my cows went to market, she weighed 755 pounds. We shipped our own stock to Chicago at that time. The weather shifted to warm and wet, but the rain fell in the night and our work was not interrupted.

The dry fall discouraged Uncle Fowler in sheep-raising and he sold the whole flock for $100. By some streak of fortune, four lambs escaped, and I got them, which started me in sheep, and I have always kept a flock, sometimes as many as fifty. I liked to raise sheep, cattle and horses but detested hogs and hens. Although there may be money in them, I didn't like their habits.

Myrtie was taking the teacher's examination. Warren was lost in the trackless jungle of politics from which he would not extricate himself until after the election. Perhaps he may be wiser if not sadder, as the political sky "looked like a storm was brewin'." School was out for a while, and Willie Howe came to sojourn with his aunt Rose.

Some snow arrived in November. That Tuesday we voted, and the result was very discouraging to me.

I got $10.68 for that cow I shipped to Chicago, so I paid on three debts. I also sold four sacks of "stuff" for $2.92 and bought 150 pounds of flour for $2.55. I worked in the corn most of the time, husking and hauling as circumstance would permit. Lewis was working at a carpenter job for Long. Our baby began to talk and trot about the floor in a most approved style.

December brought fine weather for the season. I read a history of the Jews and was also reviewing a history of the world as opportunity offered, and I was dumfounded to think how little I knew, both as regarded God and humanity.

A sort of fog hung over the hills that trimmed the trees in a Christmas garb of hoarfrost. We husked corn and hauled wood besides putting the finish on my barn. Lewis found a cowbell that looked and rang like our old pioneer bell. I cleaned it up, and discovered that it was in truth the old guide that I had followed in the 1850s over woodland and brushland in search of the cows, until Moses put it on an ox that lost it in 1863, I think, not much later than that.

My notes for the past year were rather meager, but so were my receipts. I still lived, but didn't know just how long my supply of feed would hold out. It seemed that when crops failed, my relatives multiplied and their appetites were doubly voracious. All of which was more tolerable than their abominable, soft, silly, slobbering, love-sick giggling.

Our community was thrown into a foment in December 1894 by the closing of the schoolhouse against the church people. Later, the embargo was removed to some extent. I don't get mad very easily, but when the Waukokee school board closed the school against religious meetings, I took it to heart and said and did some very foolish things. The people called a meeting and set to work to build a church. The undertaking seemed too great, but "by hook and by crook" and lots of begging and hard work, we accomplished the job.

In early 1895 a pledge paper was passed around to help build the new church at Waukokee, and I put down $5 and planned to double it if it must be, for I felt that God called for it in the interest of peace and his Holy Church. And it would be a poor faith that staggered at so small a sum, or any sum that was actually needed, if it was really God's cause.

Our church society was at last in working order. I cut and hauled blocks to serve as a foundation temporal to the building that was going up. Twenty-four by thirty-two feet was the size we proposed to build and hoped that a mild week would see a great deal done towards its erection as all hands were expected to take hold and work with a will.

Another blizzard and cold wave came on January 21st. We had lots of snow, and the wind made it a disagreeable thing to have. February continued the cold with very little done all week except chores. The lumber for the church was most all on the spot, but I had done nothing except haul stone so far. Lewis was sick, and Myrtie was taken sick much as she had been the year before. Warren went to get some medicine for her.

Lewis was still sick after a week, but Myrtie was again on the mend. After a week of anxious watching, we began to relax our vigilance a little. Dr. Phillips got the credit for breaking up the trouble and we felt everlastingly indebted to him.

February 12th was a noted day as the Waukokee Methodist Evangelical Church laid the cornerstone to its new building, and then we set to laying the other stones. Soon the frame was mostly up except the rafters. I did not help on the frame except a half day, but I stuck by the foundation until all was done from the first.

We worked on the church through a squally March, with much mud and slush. Most of the shingling was done and the chimney was up. It was a very nice-looking structure, we thought. Lewis stayed by until most of the carpenter work was done, as he was very handy in that area. I helped on the church, lathing overhead. I went to town with five sacks of grain and got $3.37, to raise money to pay my pledge to the church, and when I passed the door I saw mourning and knew that the soul of my dear friend was at last set free. Lewis and his mother went to Mrs. Frasier's funeral while I tended to business.

St. Patrick's morning was clear and fine. I put in one full day on the church roof that week, went to a town meeting one day, sawed our logs one day, and we had a blizzard one day. I put out the bees, looked them over, and united two weak colonies. I now had only two hives.

Much seeding had been done. I sowed ten acres of oats and the fields on the hill were ready. My winter wheat was dead and I planned to put oats on it. Our church was so far completed that we planned to hold a meeting there at the end of March, if we could get there in the rain. And, well, we did go, and had a nice meeting in our own church house.

The building of Waukokee Church had occupied our time all winter until spring and was the occasion of much dissension and scolding, not to mention some swearing and dissatisfaction. But on my part, all went well until after it was finished and Brother Bruce undertook to close it against singing school. Then, I said something and appealed to the pastor in charge. He read the "riot act" to Brother Bruce and we had both peace and the singing school. Now we could bid farewell to quarrels and jangles on account of the building work and hoped to go at other business, forgetting the unpleasantness of our late occupation.

All my small grain was in but the four-acre lot on the first hill east of the road. A part of the week's work consisted in a fine, large, husky, intelligent, industrious BOY appearing at Warren's, born April 2nd. Jennie and Warren named him Elvyn Cummings.

A sad day for us came as we returned from Rose Barnes's funeral. She died April 15th, of consumption, at Harmony. She left two little children and a kind, loving companion. Poor Rose, she wept bitter tears at the thought of parting but, I was told, she died resigned to God's will—and this was a comfort to me.

Myrtie went to Forestville to begin teaching a term of school. "Old Dan" got hurt while coming home from visiting one Sunday. I gave the veterinarian a dollar to come and see him, and my work was much hindered by his absence. Myrtie had been gone a week, and her Pa didn't know what she was doing, wished she'd send word, and missed her terribly.

The swallow and whippoorwill were here April 26th. But one remarkable thing was that the bluebird came not to our homes as yet. Sometimes I fancied I heard a note, but it appeared to be only a fancy. We noticed also that our stock of birds was growing less every year. This was a sad thing to contemplate at best, and there were those who could see signs of dire calamities in it to the human family.

I went to Forestville to pick Myrtle up from school. I put in corn and potatoes as usual, but decided against trying buckwheat again, as it had proven a disastrous failure for two years past. My brown cow had an ox kit.

Lewis got a hay fork from Chicago, and we were busy adjusting it for some days. This, with hoeing my potatoes occupied my spare time till haying came. We had two frosts at the end of June that did no damage. Lewis put the fork and carrier apparatus in the barn. My potatoes were hoed and looked splendid, beans and corn same, and grain was too nice to last. Frank Norton was hired to cut my grass and he did a good job of it with that new hay fork. Warren went back with Myrtie to Forestville, and Lewis went on a pleasure excursion to Clear Lake, Iowa.

My harvest was not yet done by the end of July. Its completion was a thing of uncertainty and it looked as though I should have to buy a machine. The night after my hay was gathered a blizzard knocked our grain down and that was the cause of all my trouble about my harvest work. Every machine in the county was busy and it was every man for himself. I struck out bravely with my scythe, but it was a slow job. My grain on the hill could be harvested with a machine — it was ready, and I decided I had to buy the equipment. So I purchased a Buckeye harvest machine for which I was to pay $105 over time. My grain on the hill was in shock, and so was Warren's, but the stuff on the bottom made me trouble due to wet.

The end of August found me with a sore body. My grain was stacked and we had rain again. Sunday school attendance dropped, and Brother Bruce pronounced it dead.

My threshing was done in early September, yielding 890 bushels. Wheat was at 60¢ and oats 15¢ per bushel, but my grain was so mixed that I couldn't tell how much of each there was, nor could I separate it very easily, so that it would be some time before I could tell where I was at. Lewis hauled some loose oats to town for 15–18¢ per bushel.

A very fair Fair went on in town. I was down at it two days.
Mrs. C. took five pounds of butter and a quilt, and both took a first
premium, which partially compensated her for the trouble and
money expended on the affair.

A stormy week brought much wind—one evening was almost
a cyclone. The old tree below the house went down, but it was
weak at the roots. I remember noting earlier that no bluebirds were
here, but by fall quite a number were flying round. They were very
quiet, though, and had I not seen them, I probably should not have
noticed them.

I took Myrtie to Forestville so she could go to school there and
lost our old collie along the way. She had wandered off and I
feared she wouldn't come back, as she was eleven years old and
very deaf. The previous week, a team had run over her in the
road—she didn't hear it nor know of its approach until she was
under the horse's feet. She was a good dog, but was sort of worthless by then.

November blew in six inches of snow. I sold a heifer for $14
and a hog for $10.50 to pay a note on my harvester. There was a big
earthquake all around south of us, but we didn't feel it and no one
was hurt. Frank Frasier lost his good right arm in a farming accident. I de-horned another cow, which was a serious job, for the
poor brute almost bled to death.

Warren and family were here for Thanksgiving Day. Christie
Carnegie Tinkelpaugh was visiting in our neighborhood. I don't
remember if her name appears in this book elsewhere, but if not, I
will here state that she was one of the few surviving schoolmates of
my youth. I can say I have many fond, sweet, sad memories that
her old, familiar smile and winning voice brought back with
seeming twofold force. She was to be at church, and I feared I
should be too much occupied with old memories to act the man. I
expected a poem would grow out of the harrowing my poor heart
was taking, which others would laugh at and say "poor fool."

The Ladies' Aid had a sale of articles too numerous to mention,
which brought upwards of $35. I and mine invested in over $5
worth of quilts at the auction. It was all for the preacher, and three
bed quilts came in handy that time of year.

—o—

The Problem of Spiritualism

Over the years, Fred seems to have explored aspects of spiritualism bordering on the supernatural. One of his earliest poems depicts an angel visit, a theme that he returns to in later years. He also questions Christian dogma and various interpretations of the Bible at the same time that he professes a deep faith and prefers to glean value from Bible lessons firsthand. This expresses his search for the meaning of life, for guidance during difficulties, and for answers to questions about life after death.

In looking back at the events of 1866 and 68, something was written in regard to "spirit rappings" which expressed my views at that time in as clear a light as possible with the light I then had. But there was a mysterious story that appeared in the papers of 1873:

> In the town of Newburyport, Massachusetts, is a schoolhouse which is the center of attraction. Here, a little boy comes and plays tricks with the schoolmarm by looking in through the entry window. If she tells him to go away, he only smiles very pleasantly at her and when she attempts to catch him, he seems like a cloud of smoke. He sometimes makes much noise in the attic by walking and rolling something which sounds like a cannon ball whenever the ventilator is raised, so that "Miss Schoolmarm" keeps it closed. Locks and keys are held in contempt by this mischievous little spook as he opens and shuts the bolted door with a slam, which bids defiance to the skill of men or the strength of iron. Now, this is the testimony of everyone in the school, big and little, and looks like "spirit manifestations." The teacher declares she is no believer in spirits, and is at a loss to account for the apparition.

This type of phenomenon intrigued many of us, and we explored and read much about the topic. Do we live on after our bodies die? How else could these strange apparitions be explained? Were they spirits or angels? Is a spirit an angel, or are angels divine and spirits of man?

Our neighbors were a little afraid of manifestations already developed. I even heard or saw things on occasion that were puzzling. The problem of spiritualism remained unsolved for us for many years, and we continued to investigate it a little.

Spiritualism also attracted many scientific minds. Some very remarkable manifestations were witnessed. A lady lectured in Preston occasionally who was said to be in a trance. She held live coals in her hands without sustaining the slightest injury.

For more than seven years I had been investigating spiritualism with constant, perplexing disappointments; almost grasping the mystery, and then finding it a shadow. But, still believing a shadow as evidence of an object, I renewed my exertions and determined to admit no evidence until it had been tested, pro and con. Then I caved and was ready to accept the theory and say: I believe if a man dies he shall live again. This was my conclusion about spirit communion, that our departed friends could visit and commune with us under favorable circumstances, but it was detrimental to the health of the person through whom they reached our sphere of understanding. My own experience in this matter tended to make me cautious about how I exposed my health. Once satisfied that man was immortal, we could afford to wait, rejoicing in hope.

How strange it seemed that one day one was a living, moving, thinking being and the next, an inert lump, changed in a moment to merely nothing that the wondering world knew of. For when some spiritualist dared make the proposition that we lived on in spite of nature's changes, he was called a lunatic, or at least it was said he was laboring under a hallucination and a delusion.

A very curious spirit phenomenon was reported as having transpired at Owatonna, Minnesota. A Mr. Diment was so annoyed by unseen beings that he was obliged to leave his house. This may be explicitly relied upon as true, but I believed that a proper course would have relieved him of his unwelcome visitors. This running away and raising the cry of devil would never satisfy anyone, either mortal or spirit. Almost anyone would be civil if treated civilly and I believed that a firm, manly, honest investigation was all that was necessary to find out the cause of disquiet in spirit phenomena, and bring about order and peace.

Then we read about another strange item, but couldn't really believe one word of it: a spirit medium was carried 150 miles without any visible means of support from Chicago to Madison, Wisconsin. We learned later that this was a put-up job, and this type of charletanism made it difficult to know what to trust or accept as fact.

In looking back, this was about the time that saw the beginning of the end of spiritualism with me. I saw more and investigated with greater thoroughness than ever before, and must admit that many things happened to me and under my observation which to this day I cannot account for. But fraud and false pretense were so much and so often in evidence that I bolted the business of depending on others for testimonials. Then, like a chain, it proved no stronger than its weakest link, and fraud and lying were weak spots in a very plausible and beautiful theory. Thus, it occurred to me that if there really were spiritual beings desiring to commune with mortal man, they could best serve that purpose by communing to me alone. For where two or three were gathered together there was too much chance for skullduggery, and upon this hypothesis I concluded to rest my case. It then had a good long rest, and all was quiet. No disturbance of any kind. No communications from the spirit world in any tangible shape or form, and so I gathered that if spirits there be, and there may be, they had no particular business to transact at my house.

Spiritualism seemed focused on teaching thoughts of immortality, which hope was not the fault of the sect. The source from whence it came was where the error lay. Mankind had no right to claim it—if there was a life for us beyond the grave, it must be a gift of divine grace. And we had nothing to prove this but the Bible, hence all other claims or pretensions to a knowledge of future life are, of necessity, false and must end in flat failure. What the Bible *really* teaches us is a question upon which men differ so much that I thought I must content myself to let the discussion entirely alone.

Therefore, my advice to the young in regard to spiritualism is: don't monkey with it. If you have time to spare, use it to a better purpose. Study science of a better sort, learn geography, Latin, Greek—anything—study the Bible or botany, but leave spookism to cranks and idlers. As for me, I decided that all the spiritual knowledge I shall ever receive must come from God's own Word, so I took every opportunity to make myself familiar with it.

Love of debate in my younger days carried me into many unnecessary controversies that were sources of deep regret in after years, especially religious debate with the preachers at Waukokee. In looking back to when Rev. Sheets and Rev. William Hipes of the Brethren Church lived here, I wrote several letters that would have been better left unrecorded. I don't much doubt that I had the best ground for argument in both instances, but I learned later to listen in silence whether I agreed or not. In looking over the record of that year I am moved to express regret that I should engage in a controversy where neither party had the least true or real knowledge of what they were talking about. Preachers do what they can to explain the "Mystery of God," but at best they are naught but hard-working, fallible mortals, stumbling through ages of mistakes.

One year, for excitement, a Universalist preacher visited Waukokee once every month and stirred up the locals with argument and logic about our sinning ways that seemed unanswerable but was not pleasant to hear. Only those who were ignorant of scripture made any attempt to stop him and, of course, their quotations were too ridiculous to mention. One lady quoted the "old saw" about the tree that falls so shall it lie, and added as death leaves us so the Judgment finds us. The preacher disputed the quotation and asked her to find it in holy writ when, lo, she did not know whether it was in Genesis or Revelations, but contended it was there. I heard a Baptist minister who went regularly, remark "I am enough for that man. I think his preaching is false, but he will never know it by my telling."

I held the theory that any system of religion that tends to create in us a hungering and thirsting after righteousness, and which elevates the standard of morals in a community is a true religion, and anything short of this is false. If I am right, the scripture test would be "By their fruit ye shall know them," and argument on

points of doctrine would dissolve into a very simple problem —
what dost thou work?

The argument of infidels that all prayer is useless seemed to
have taken partial possession of me in 1877 when I wrote the
"Grasshopper" poem. This was in answer to Governor John S.
Pillsbury's proclamation setting apart a statewide day of fasting
and prayer because of the plague of the locusts on our western
frontier. If any confession is needed on the point, the poem itself
will furnish it.

The Grasshopper

In love and wisdom God has made
These insects, pests, and sent them here
Upon our fertile fields to raid
In fell destruction year by year.

For our correction, we are told,
This dreadful scourge on us has come;
Our children die of want and cold
For deeds that we ourselves have done.

The wicked and the righteous dwell
In close communion side by side,
A common fate awaits them all
Destruction dire, far and wide.

Thus age on age has rolled away,
Men doubt while looking o'er the past
For ever when the Lord doth slay,
The wicked suffer with the just.

Our prayers ascend, "Why, Father, why
Do plagues and storms and droughts assail
The man whose thoughts and aims are high,
If 'tis for sin these things prevail?"

In true humility we pray
None may thy Holy Will oppose,
But when thine armies rise to slay—
In mercy, turn them on thy *foes*.

I am now ready to say, however, that "where I was once blind, I now see." I do think people should pray, not so much for temporal blessings as to open up communication with heaven and draw nearer to God. Any further comment of the subject is probably unnecessary. I pause to mention another poem of that year's production, "The Captive Sheep." The circumstances mentioned in the poem are partly true, but mostly fiction. But the devil theory as generally set forth by preachers and teachers of theology never had much sympathy with me and doesn't yet; or, perhaps I should say I never had much sympathy with the theory. I can't believe such an evil entity ever was created by an all-wise being, and the stories we read about him in Scripture are very obscure, not to say contradictory, and smack strongly of mythology. I am sorry to break in upon a doctrine so dear to superstition's devotees, but a little jarring does the world a lot of good sometimes, and I promise some more of the same treatment later on. The principle of evil has no doubt existed ever since there was a rational being to act and to choose between right and wrong, and I suspect the one who chooses or acts the wrong may justifiably be called a devil, or doer of evil.

The Captive Sheep

My father had a hundred sheep,
One hundred sheep and one;
Ninety and nine as free as air
Upon the hillside run.
But two, the nicest sheep we had, were in the barn confined.
      I, sorely puzzled why it was
      That they in jail should lay,
      Inquired of Father for a cause
      Upon one Sabbath day.

Ah, little do you know, my child,
What these two sheep might do;
Though now they seem so tame and mild
Turned loose, they'd murder you.
So to their prison life, my lad, these sheep must be resigned.
  He prepared my lesson then to say,
  As I stood beside his chair,
  Such was my wont, and on that day
  It fairly raised my hair.

For then he told me frightful tales
Of Satan and his power;
How he, like roaring beast assails
God's creatures to devour,
To spring on unsuspecting man at every crook and turn.
  Then trembling, pale and sore afraid
  I felt my days were few and evil
  When a thought came to my aid,
  "Father," I asked — I never was uncivil —

"Our heavenly Father is so good,
He hears the ravens call,
He gives the little sparrows food
And counts them as they fall,"
But then the attitude I voiced gave Father much concern,
  "His voice sounds o'er the angry main
  And lulls the waves to sleep;
  Why don't he then, his Devil chain,
  As we do our old sheep?"

In 1878 I took an interest in a revival meeting conducted by one
Rev. Damon — a stranger styling himself a free Methodist, but his
theology did not strike me as altogether rational, although he
seemed very much in earnest. Crazy people often get very much in
earnest on religious topics, but oh, what wild theories they
advance! He was that way, I thought.

I confess to the belief at that time that men must and could save
themselves by their own good works. I lived in hope and expecta-

tion of a better and more glorious life beyond this present state of things for those who aspired to it, as I believed our works would follow us. This was nature's teaching as well as the plain teaching of common sense; and considering it thus, who of common sense and ordinary human integrity would not act as becomes a rational being? And acting thus, it puzzled my mind to wonder what need there was of a mediator between God and man, since acting irrationally brought its own condemnation.

I am sorry to have to own up to this folly, but excuse myself on the ground that there were others who shared this approach. Dr. Ross, for instance, tried to soothe his tortured soul by reasoning thus: men will be judged like a debit and credit account in bookkeeping; if the good deeds overbalanced the bad, the debt was paid. A very captivating theory to be sure, but altogether unscriptural, and sure to mislead and ruin those who are deceived thereby. *Good works are commendable, but they cannot compensate for sin.*

My religious conviction seems to have simmered down to a belief that the religion that prompts mankind to behave themselves was the genuine article. With such a creed, no man need fear to die and without it, any rational being ought to fear to live.

All this churned around in my head until 1887, which brought about a decided change in me as to religion. There was little to record, however, except that while working alone at my plowing and thinking over my past foolish wanderings, I began trying to untangle the problem of life that I had, by my egoistic folly, got into such sweet shape. Finally, I gave up in despair, and was constrained to "give up myself" and call upon God to save me from my skeptical vanity. Then, what a flood of conviction came rolling in upon me. All the vanity and sin of twenty years rose up before me and my spirit was gone out of me. I feel sure today that divine love came to my rescue and saved my soul, otherwise, I should have turned back again to wallowing in the mire of self-conceit and unbelief. Just how long this stage of contrition and repentance lasted, I don't know, but the chronic form which it took on stays by me to this day.

I shall never cease to regret past mistakes or pine for lost opportunities. Twenty years' wasted time, I call it, but perhaps I am too severe a critic and, possibly, the lesson had to be learned. This is the one crumb of comfort I have to console me when I get blue over past follies. I know I am not what I might have been but this, too, cannot be changed by pining and I call it a waste of present, precious time to do so. So, forgetting the things of the past, I press forward toward the mark of the prize of my high calling. I write the above knowing full well that religious humbugs are as numerous as any and all other swindlers put together.

After that, my religious experiences of 1892 were the next most significant of any year. It was said that Abraham Lincoln withstood the importunities of the Abolition Party to free the slaves until they quit him in disgust and despair. But when Lee's army crossed the Potomac and began its devastating march north, he realized his dependence on God and, in doing so, became aware of the fact that it was necessary to get right with God; hence, his emancipation proclamation. This, to some, may appear a cunningly devised fable, but the fact remains that our Union was saved from that very hour and our extremity of affliction seemed to be God's opportunity to console and heal as well as to convince and convert.

Much has been said and written about answer to prayer, and some prayers are apparently trivial and foolish in proportion, no doubt, to the amount of mental capacity possessed by the writer. Be this as it may, I gladly confess that I was at my "wit's end" at the time that Rose was stricken by cancer, and the Lord heard my cry and helped me out of all my affliction. Not only that, but I hoped it thoroughly cured me of being an infidel, with all its various forms and phases with their plausible arguments, which at best only show the ingenuity of men.

I was glad that I had already come to these conclusions and had these experiences by the time the new century came. I was on sound footing, but those who were still searching became fuel and fodder for all manner of cranks and false prophets that held sway with predictions of doom on the advent of the new millennium. People even searched the Bible for proof of the portents of Armageddon, reading all kinds of meaning into the annual bouts of cyclones and earthquakes that affected the earth. It was my turn to be a humbug to those who put stock in such preachings. The seasons and cycles did not seem any better or worse to me than those of previous years, and I determined to keep toiling and living as I always had done.

— o —

Warren

The journal covering 1898 through 1902 is missing. We may never know for sure whether a volume was lost by a family member or whether Fred himself destroyed this volume late in life when he burned "some useless papers and letters." During this period, Uncle Fowler dies and first Lewis then Myrtle move far away. This would have been a painful and difficult time for Fred, so it is conceivable that he destroyed the journal, especially if he had noted his disapproval or sadness about his children's choices. Luckily, I had other materials to help fill the gap, and am focusing the next three chapters on each of the children. These chapters are written by me, though I've tried to replicate Fred's voice. The source materials for Warren's chapter are found in a huge scrapbook of clippings where he pasted poems, editorials, cartoons, and articles. Excerpts are not foot-noted, as the clippings do not include the dates or sources. The typeface suggests that most of the articles are from the Preston Republican, The Courier, *and* The Toledo Blade. *Some of the articles or letters to the editor are by Warren, while others represent subject matter that inter-ested him. As a teacher, one of Warren's main concerns is the quality of the educational system, and it is fascinating how many of his opinions are still expressed today by contemporary educators.*

At this point in time our children had come of age and were growing in independence. This was a source of both pride and pain to their mother and me. Pride as we watched their accom-plishments, and pain as we suffered anxiety over decisions made and distances that separated us. In early 1897 Myrtle was twenty years old, Lewis was twenty-four, and Warren was twenty-seven. Uncle Fowler died that year, leaving me the homestead farm. He took me in as a young boy, raised me as his own, and taught me the value of hard work.

Warren and his family lived at Fountain, but they returned here for a spell to assist me in overcoming the loss and inconvenience sustained when Uncle Fowler's old barn burned. In 1900, Warren was elected Assessor of Carimona Township and moved his family to Isinours, five miles away. We usually saw them at least once a week, and took great delight in watching their children grow. We exchanged work, as Warren helped me on the old homestead and I helped him farm and cut wood on his acreage.

Even though he was teaching, much of his time was spent attending political meetings throughout the county, and writing for the newspaper. These articles were a forum for Warren to express his views, but I didn't always agree with his conclusions. For the most part, I chose not to engage him in debate and tried to keep my mouth shut unless I feared he was being swayed by undesirables or smooth-talkers who were not sincere.

Warren's political views pitted him—allied with the Populists, or People's Party, and Ignatius Donnelly—against Major Hotchkiss and Martin Kingsbury's paper. Their editorial columns and letters accused each other back and forth and back again of all manner of corruption, stupidity, and duplicity. Major Hotchkiss wrote one week that A. L. Long, Ignatius Donnelly, Warren Cummings and a host of other fellows sacrificed many of the best years of their lives "sawing wood without compensation, dipping their literary quills in fiery ink and doing a little word building intended to be detrimental to the oldest living political and religious curiosity now extant." Donnelly replied thus:

> Poor old Hotchkiss! He should have remembered all the gattling guns which thundered against us in vain for thirty years past, directed by the giants of plutocracy, and not attempted our destruction by the insignificant little potato-wad popgun which he wields in the abysmal depth of Preston. We never see Hotch's obscure sheet, and don't want to see it. It is too much like the Yankee's definition of nothing—the small end of a small stick, whittled down to a point and the point cut off! May a merciful God remove the old man soon to another world and not permit him to sink to any lower depths of baseness and degradation in this.

Ignatius Donnelly was a notable figure in this world. He was first elected to Congress in 1863, continued in politics for many years, and helped found the People's Party. Discoverer of the cryptogram, he proposed the theory that Francis Bacon wrote Shakespeare's works. Warren met and corresponded with this great man often during this time, in the sunset of Donnelly's life.

When Ignatius Donnelly died, he was lauded for his distinguished scholarly and political career which started out as Lieutenant Governor, then he was elected to Congress for four terms, and he was a great reform leader throughout. He was unparalleled as a humorist and an orator. But some enemies called his a life of ambitious agitation, wasted energies, and lost opportunities that ended in a string of failures. He was permitted to see the first day of the twentieth century and, at midnight, at the end of that day his earthly career ended.

Warren and Tom Meighen spoke at a meeting of the Preston Silver Club, and they showed France as an example of successful bimetallism while England and Germany were still under the curse of gold. They constantly and consistently argued that international free trade and tax policies were in the best interests of the trusts, which is to say, not in the best interests of farmers. Farmers sold to the trusts who set prices for the farmers, and then the trusts benefit from free trade abroad, not the farmers. "Oh, stupendous and villainous duplicity of this plutocratic political power which drives our own people out of honest industry and into soup houses, saloons, and the mouth of hell and then turns about to pacify you with false representations of tariff legislation in your favor." Warren agreed with one J. Doran, that no matter how cheap the clothes bought from abroad, they come at a very high price when taking into consideration the fact that we leave our own skilled and unskilled labor unemployed and our good raw material unused. No more fusion with the devil or the modern Free Silver Democrat; equal rights to all men, special privileges to none, and middle of the road men at the helm.

When the Populist congressional convention was held, the press reported that Warren Cummings was elected temporary Secretary of the Convention. On motion, a report of the Free Silver conference was heard. They selected the Honorable Patrick Fitz-patrick of Winona as candidate for Congress. Thomas J. Meighen addressed the chair and, on behalf of Fillmore County, seconded the nomination. The candidate then made an eloquent acceptance speech. The party urged people to vote for Fitzpatrick to succeed Mr. Tawney, the gold bug champion.

In 1898 Warren waffled for and against John Lind's candidacy for governor. People should be permitted to change their minds, but it is difficult when one's opinion has been printed and publi-cized in the press. Of course, it was his own hand that loaded that gun.

That was the same year that this community was greatly agitated over a bank failure. Widows, washerwomen, sewing women, and many other poor people lost their all, and the total of such losses reached up into the thousands. The banker was one Mr. Todd, a regent of the state university. He was a very agreeable fellow to meet, had been mayor of Preston, a member of the school board, and treasurer of a church organization. In appearance and personal conduct he would compare very favorably with other regents of the state university. As Byron said of the pirate: "He was as mild a mannered man as ever scuttled ship or cut a throat." But oh, the devastation he wrought. God pity the poor victims.

In 1902 Warren's long-time friend and political colleague, Thomas J. Meighen, was presented as the People's nominee for governor. He lived in Forestville,[*] about seven miles from here, and Warren spent many days and evenings there engaged in debates, writing and making plans. Even before the nomination, Warren worked tirelessly on Meighen's behalf, and he was a dedi-cated campaigner throughout. The following is an excerpt of Meighen's gubernatorial platform published in the paper:

> The People's Party represents the electorate's sovereign right. It opposes granting special privileges to any class or corporation where the people's rights may be jeopardized. It is against private ownership of utilities, and against trusts controlling

---

[*]   Forestville is now a historic site managed by the Minnesota Historical Society. The Society's archives include correspondence between Meighen and Warren Cummings.

prices or goods or movement of goods. Trusts are the result of private ownership and control of the three great instruments of commerce: money, transportation, and transmission of information. The one remedy to the trusts is that ownership and control be assumed and exercised by the people.

The people's flag of distress is up in the form of thousands of miners and their families who are at the point of starvation. The whole business of the country is about to stop for want of coal, one of nature's resources put on this planet for use by man. Yet the people's flag of distress is not heeded by Washington.

The best answer is for the coal mines and other public utilities to become property of the people, to be operated and managed in the public interest not according to private trusts. What the farmers and merchants need is relief from the high prices of transporting their goods. If elected, I promise to enforce the law against all merger trusts and other violators. Our party is pledged to reduce freight rates upon the railroads of Minnesota.

The People's Party was established to safeguard every sacred right of every citizen. This is true not only of the state ticket but of every candidate of the People's Party. If the voters of Minnesota will investigate these issues before registering their verdict, we will win a glorious victory.

Warren's own letter of support ran thus:

I have followed the Populist flag since it was first hoisted, attended two state conventions and all our county conventions, and always read our papers. I find the name of Thomas J. Meighen connected with every reform movement since the days of Peter Cooper. He was in the fight before he could vote— declared then that he had "enlisted for the war" — and his career has sustained that assertion. He is not a candidate for personal gain, he does not need it, his name here in Fillmore County is synonymous with success in any line of business. He offers to be governor of the whole people, and his services on Lind's Board of Equalization proves his superior qualifications for the place and his uncompromising fealty to our cardinal doctrine: "Equal rights to all, special privileges to none."

He has sound ideas for running the state twine plant and, if given the opportunity, would break the back of the book trust with state manufacture and free distribution of school books. The

Owens and Bowlers may have tried to be populists, but it appears that when men have grown gray in the services of a party, they are ever too prone to follow the bell wether with that party tag in his ear. A big vote for the People's Party ticket in Minnesota this year will worry plutocracy more than a dozen victories for a mixed organization of old line and new line democrats.

Despite the hard work of our area Populists, Samuel R. Van Sant was re-elected as governor of Minnesota.

***

Another topic that Warren addressed frequently, maybe even more so than politics, was education. As a teacher with fourteen years of experience, he was concerned about the rules and regulations imposed on him by so-called state leaders who didn't know the realities of teaching in a rural school. For example, on the topic of scarcity of teachers, he wrote that there were two possible explanations for the scarcity: 1) That teachers were underpaid, and many left for other professions that offer better wages. 2) That new requirements for certification exclude country-taught teachers who may lack certain academic qualifications but have experience, while younger teachers use the profession as "stepping stones" to other things, leaving after a short while.

Warren was a critic of the "summer schools" held for teachers. He encouraged use of summer school reserve funds to be used toward the purchase of a proper library. More could be accomplished through access to more books than by teachers deserting their homes in the harvest month to spend their last dollar and wear our their nerves in the sweltering heat of a summer school. Access to more books would benefit the students by helping them acquire the habit of continuity of attention, as the mind that flits from one subject to another will not absorb much. He asked readers if their reading habits, and those they teach their children, encourage knowledge or consist only of reading bits in newspapers. He maintained that the mind of the modern boy or girl is chock full of "science" yet almost wholly ignorant of the three "R's," which are so necessary in the practical affairs of life. Additionally, the introduction of the "vertical system" of penmanship should be deplored, as penmanship seemed to be one of the lost arts in the schools of the day.

He agreed with one Miss Novella M. Close, principal of a public school in Chicago, who stated that education begins at home; that early education and spiritual development are critical to children's success in school and in later life.

Practical, not theoretical, education is what was needed in the rural schools. This Warren learned from his own experience and from reading works by Lord Bacon, who he maintained was one of the most enlightened educators that ever appeared on the earth. As the father of inductive philosophy, he showed us useful and practical ways to educate. Inductive methods could be more useful than the platonic philosophy which had infested our systems of education. Bacon taught that geometry and other branches of mathematics was valuable as a branch of education only so far as it contributed to supply the wants of society.

Warren grew indignant when one Miss Anna Holland taught school in District 110, and the district refused to pay her. She sought assistance from the courts and won. Olmsted County officials claimed, after the fact, that her certificate was valid in Winona County, but not Olmsted, yet they had been the ones to hire her in the first place! Rightfully, the court cared more about labor performed than certificates granted or withheld by the whim of a county superintendent.

The National Republican was foremost among the press of the state in promoting school interests. One of the best articles was about the importance of discipline, which Warren heartily agreed with and clipped for his scrapbook:

> The success of a teacher in a public school depends neither upon his scholarship nor his ability to impart instruction so much as upon his disciplinary power; his capacity to "manage" a classroom. Fine, knowledgeable teachers may be lost if they are indifferent disciplinarians. Too much time of many teachers is consumed in efforts to persuade or dispel a few mischievous or vicious pupils to conduct themselves with sufficient degree of propriety to permit recitation to proceed. A few youthful rowdies are permitted to deprive a school of the services of an able teacher.

While it is true that teaching is an exhausting profession, it is a fallacy that teaching is exhausting. The health and temper of teachers are wrecked not by teaching but by governing. Our schools are not schools so much as nurseries and reformatories. The proper work of the teacher—instruction—should not be made secondary to the duties of the parent, who should teach behavior, or the constable, who is the public disciplinarian. The boy who disrupts the class should not be permitted to remain. Summary dismissal should be the school's action. The payment of taxes confers upon no man's child the right to deprive another man's child the right to the instruction of the teacher. Reformation of the disruptive pupil should be affected by the parent or civil officer before such a child is permitted to associate with those receiving instruction. The insolence and want of respect to authority which is a marked characteristic of young America is certainly stimulated by this defect in his training.

At times Warren felt that people were not paying attention to his well-reasoned commentaries. He began one of his letters to the paper in this way:

Will you kindly give space to the following extracts gleaned from the pages of the State Educational Association report? I make this request because the opinions set forth in these paragraphs are in accordance with views which I have from time to time expressed through your columns and, as the authors receive more corpulent salaries than your humble correspondent, their testimony will of course possess peculiar force.

The article went on to say that the principles underlying all good teaching are exemplified by the teacher in the little red school house as much as by professors of pedagogy in our more pretentious schools and colleges. That child study is too apt to fall into the line of the study of the hypothetical child, and may in that way lead to false conclusions. Many a time in the country school house, the uncultivated but inspired young girl has shown a knowledge of the workings of the child mind that would put to shame the theories of the most learned and profound philosopher. The multitude of studies required in the brief time allowed in most of the schools is ruinous to the very aim in view. It leads to confusion of ideas rather than clearness of thought. Success of the old education was making the work intensive rather than extensive.

The paper generally supported Warren's views, which is probably why they printed them so often. Once they replied:

> Mr. Cummings has communicated on the issue that there is much in our educational system that needs the pruning hook of common sense: some of the text books should be discarded, some of the methods in the school room need reforming. In this necessary work it is proper for teachers to take the initiative. We regard Mr. Cummings one of our best common school teachers. His contention against the summer schools is proper. Some of the the instructors could not successfully teach a district school. Their having got through a normal school humbug with a diploma is no evidence of qualification. What should be done is the creation of a professorship in each county capable of guiding teachers up to the highest standard in school room work.

Education and politics came together during the 1902 election, and *The Minnesota Forum* printed Warren and Tom Meighen's opinion on the educational plank of the People's Party, which they had helped draft:

> The state superintendent raised an outcry over the People's Party plank on education reform. It all depends on what the aim of education should be. The current system is designed so that the common schools are feeders for the high schools, and the latter exist only for the purpose of preparing young people for the university. If that is the true aim of the system, the present management is right, but we think that the high school and the university should not be the aim of the common schools. The great majority of pupils in these schools never finish even the common school curriculum, why then, should they be hampered by studies designed for the few who aim to reach the higher schools?

> Why should not the studies be so arranged that in the few short terms that the average child can attend school he can get a fair amount of knowledge of those things most important for him to know as a manager of his business and as a citizen? Why should he, for instance, be obliged to study physiology for several terms, when all the knowledge of that subject necessary for him to know in order to lead a hygienic life can be given in a dozen lessons?

The common schools should be divorced from the university system. The aim should be to give the pupils as much knowledge of the common branches as can be given in a short time. There should be no unvarying course of study, but it should be left to the discretion of the teacher, acting in unison with the parents and the board. State aid for common schools should be made dependent on knowledge of the subjects taught, not on knowledge of things the scholar has no use for. If the subsidy were made dependent on experience, there would be more teachers who could remain in the ranks and avail themselves of the longer terms and increased pay. It is an injustice to the rural schools that they should be made by law a feeding ground for high school graduates.

Warren kept a big, oversized scrapbook of news clipping that he would show me from time to time. Many of the bits were his or his friends' articles on political or educational topics. Each page also included a poem or two that he found interesting, funny, profound, or melancholy. There must have been more than one hundred poems in that scrapbook, and we would sometimes read them aloud to appreciate the full meaning of the art. Some of our favorites were by Ella Wheeler Wilcox, James Whitcomb Riley, Eugene Field, Longfellow, George Eliot, Shakespeare, William Cullen Bryant, A. J. Gault and others.

The scrapbook also contained information on faraway lands: ancient ruins in central Asia, the dead city of Merv with the tomb of the Sultan Sandjar; Norway, the land of the midnight sun; the scenic miracles of the Yellowstone; Australia and New Zealand; agriculture of China. All fantastic descriptions of places we could only visit in our minds.

As it turned out, it was Lewis and Myrtle who ventured out to satisfy their wanderlust, not being content to stay at home and hearth in Waukokee.

— 0 —

Lewis

The primary source for Lewis's chapter is a five-year diary in which he jotted five lines each day spanning March of 1900 through 1904. The most fascinating part of the diary describes the ordeals of establishing a homestead claim near Lake Itasca, nearly 300 miles north of Waukokee. His claim has long since been absorbed into Itasca State Park, but at the turn of the last century, the land was available to anyone hardy enough to clear the land, build a home, and survive the harsh winters. Myrtle, too, goes north, and her letters home help describe conditions further.

From almost the time he was able to help with harvest, Lewis had frequently spent parts of the summer with Rose's family in Dexter. Then he and his cousin went exploring property in Iowa, South Dakota, and northern Minnesota. During these travels, he met up with the Maw family of Owatonna, and became charmed by their daughter Florence. Florence's brother, Fred, convinced Lewis to stake a claim up by Lake Itasca, the headwaters of the Mississippi River.

In early 1900, as Lewis prepared for his move, he built a cupboard with drawers for Rose out of old shipping boxes, ordering the door hinges and drawer pulls from Montgomery Wards of Chicago. He helped me with other projects such as lathing and papering Myrtle's room, cutting cord wood and building a fence, but it was clear his mind was on other things. He even lined and covered the old buggy top and re-made the seat cushion to prepare for a visit from Florence in late April.

Once Florence arrived, we hardly saw the young folks. They were off to see the flood damage at Willow Creek after Sunday school, then they went to Preston and Carimona looking at scenery, picnicking under the bluffs by the railroad bridge. One day they climbed to the hill above the Tinklepaugh spring and

then spent an evening playing music and singing. Lewis and
Florence next went off to see picturesque Lanesboro, saying they
had forded the river three times upon their return. But soon the
frivolity was over and his trunks and tool chest were packed. They
departed for the train at Isinours on May 18th, which left around
ten in the morning, scheduled to arrive at Austin by noon, and
Owatonna at four.

Lewis and Florence spent a few weeks with Florence's family
and the Owatonna cousins and left for Park Rapids on June 5th.
Florence and her mother went with Lewis so they could visit with
Fred and his family. Lewis helped saw wood and clear land for
Fred in exchange for his help later when Lewis was ready to build
a shanty on his claim. They caught forty fish in one afternoon on
Lake Itasca, but were not always lucky with hunting. One morning
Lewis had to settle for a porcupine for breakfast, as other game
could not be found. Florence left for Park Rapids at the end of June,
and Lewis was alone to start work on the claim.

Letters from Lewis came regularly, so we were able to follow
his progress. He hired on for $2.50 a day to do carpentry work on
the schoolhouse at the Scandinavian settlement and, after that,
went to work for a lumbering crew building lookout towers on
Schoolcraft Hill, around Bemidji, and at Marquette Lake. He said
the mosquitos were bad, and fleas kept one lively nights. In
August he stayed with Fred Maw and added a proper roof and
door to the shanty on his claim, but the next month he was back at
building towers, lumbering and hauling logs to the sawmill.

Fred shot a buck in November, just in time to put some meat
up for his family that winter. Lewis wrote that he had found much
more pine on his land than expected, and good drainage for the
hay slough, plus two brooks that flowed into the south end of the
lake. He was looking forward to improving his place the next year,
but it wasn't fit to stay the winter there. After spending Thanksgiving eating oysters with Florence and her folks, Lewis came
home.

He spent that winter helping with chores, cutting wood, and
taking violin lessons from Warren. We read several books by
Byron: *Cain, Heaven and Earth,* and *Deformed Transformed.* They
were very interesting, and the source of many evening discussions.

Lewis sold his farm to George Oleson for $1,000, paid off his grandfather, and released me from my co-signing obligation on his land debt.

In March, 1901, Lewis left again for Owatonna for a visit before going back to Itasca. During the layover in St. Paul he spent an evening with Florence's cousin, Mr. Leslie McCormick, and one Miss Winger at Hamline University. Once he reached Park Rapids and made the trek to Itasca, he wrote back to Myrtle that he had arranged for her board at Fred Maw's at the rate of $35 per month while she taught school and that they had picked out a claim for her to homestead.

This was no easy way to set up housekeeping. A homestead claim required five to seven years of residence, and the land was covered with trees and brush that had to be cleared. Even after the trees were cut to build a cabin, the ground would still be covered with stumps and roots, which had to be cleared in order to farm. There were few roads up there, and it could take hours or an overnight trip to the settlement or town to get provisions. Lewis being up there was one thing — Myrtle was another matter — but her strong will prevailed, though it broke her mother's heart and mine too.

Her first letter brought no comfort, as she told us that she and Lewis got the rig stuck in quicksand on the way from Park Rapids to Squaw Lake. They had to unload the wagon, leave their supplies there and ride horseback to a neighbor's house, then return to retrieve the wagon and the load next day when the frozen road made traveling easier.

Jimmy Whitney and George Heinzelman helped Lewis make a stable and put tar paper on his house roof. The job of shingling took some time. They burned brush at Lewis's claim and marked and cleared Myrtle's land. Mosquitos were bad, sometimes waking them up at four, at which time one might as well stay up and start work. Lewis helped Martin Heinzelman, George's father, clear land in exchange for help clearing his with the team. They and Fred Maw were about Lewis's closest neighbors, being an hour and a half away. Lewis cleared one acre for oats and also planted potatoes, beans, beets, sweet corn, and yellow corn. Strawberries were good that year.

For fun and socializing, the youngsters took part in a gun club, though their scores were only middling and sometimes Myrtle got the best of Lewis. Sunday school, of course, was always in order when time and weather permitted. Usually, if they left the house before eight, they could reach the settlement by ten-thirty in time for service at eleven; lunch there, and leave by two getting back around four in the afternoon. Lewis ordered some furniture for his shanty from Sears and Roebuck, which came in late June, just before he and Myrtle departed for Owatonna.

Rose and Myrna were on hand for the nuptials, I stayed home to take care of the stock. The newspaper gave a full description of the wedding:

> On Tuesday, July 2nd, at high noon, at the residence of the bride's mother, Florence L. Maw of Merton was married to Lewis S. Cummings of Preston. Reverend George L. Reynolds, an uncle of the bride, performed the ceremony. Alice McCormick, a cousin of the bride, played the wedding march. Miss Myrtle Cummings, sister of the groom, was bride's maid, and Leslie McCormick, cousin of the bride acted as best man. The house was handsomely decorated for the occasion with ferns and flowering plants. A magnificent arch of oak leaves with a beautiful evergreen bell above and in front, and a stately pyramid of ferns in the background formed a fairy-like alcove where the bridal party stood for the ceremony. After the ceremony and congratulations, a sumptuous dinner was served and a general good time followed. Fifty guests were present. Wedding presents were numerous, elegant, and valuable. The bride has many friends in this city, having pursued a course at Pillsbury Academy, and the groom is also well and favorably known here. Mr. and Mrs. Cummings departed July 4th for Vern, Beltrami county, where they will make their home. They have the best wishes of many friends.[*]

Rose brought back some odd "pitcher plants" that Lewis had collected up north. The instructions for their care were to keep them very wet, for they grew in the swamps. Lewis said to put water in the pitchers and some others said that if the plant was properly watered the pitchers will draw the water up from the

---

[*] *Preston Republican*, July, 1901.

roots and so fill themselves. But when they are out of doors in their natural state, of course, they get filled with water when it rains and I presume they draw water from the roots too.

News from up north said that the newlyweds and Myrtle all arrived well, but experienced a wagon breakdown after Mantrap Lake, six miles from the park house, which was still about two miles from the Heinzelmans' and about four from Fred Maw's place. They had to leave the load and walk three hours to the Heinzelmans', arriving at midnight, and it took them another day to replace the wagon tongue. Myrtle had left her straw sailor hat in the wagon all night and the boys found it sadly chewed on the edge of the brim in the back. On investigation, they found the offender—a young woodchuck—which they captured and put into a sack, and George Heinzelman adopted the animal. He was christened "Chuckie."

They stayed at the park ranger's house and went on to Fred's the next forenoon once the wagon was fixed. Then, on the way to Lewis's claim, the buckskin mare acted sick and couldn't pull. It was one hundred degrees, and Nellie wouldn't eat all the next day. Lewis stayed up with her, keeping a smudge pot going all night, but she died at dawn. He had her skinned out and buried before breakfast so he could get to the hay cutting and potato hoeing. He helped Florence rearrange the new furniture, and they picked raspberries and blueberries that she baked up into bread, pies, and doughnuts.

Myrtle was teaching school again, though she said the class was small as the bigger boys weren't coming any more that summer on account of work. So her work was easier, and she looked forward to having a month for sewing and getting ready for fall. Myrtle wrote:

Dear Father and Mother,

Friday, immediately upon dismissal of school, George informed me that my cabin was nearly ready to receive me so we "struck out" for Squaw Lake as soon as we could get started. Took a lunch with us and stopped for lunch by moonlight on a fallen Norway pine. While we were eating we could hear a deer moving around down in a slough just a few rods from us. He, too, was eating his supper. We couldn't see him and, although we crept cautiously and carefully as near as possible, he was a minus quantity before we were there.

Well, we reached Lewis's just as he was finishing his supper. We were tired some but not much and George declared that I walked so fast it made him tired to keep up.

The next morning we all went over to my claim where we scared up some partridges and started off in search of them, and finally crept up on one and I shot her. Then we went back to "camp" where Florence was paring potatoes for dinner and got our meat ready and had a fine dinner of partridge, potatoes, graham bread, baked beans and jelly. After dinner the boys put the finishing touches on the cabin, put in the stove, and made a fancy bedstead across one end of it of poles, tag-alder bushes and hay. Then we filled a tick with hay and proceeded to make up our beds.

The shanty we stayed in on my claim was only a temporary one of old logs. This week they are going to build a better one with a good roof and a floor, as last week we couldn't get lumber over.

Florence found an Indian's toboggan made of an oak board over near my lake. I'm going to save it. My claim is a valuable one, having lots of pine, a fine lake and lots of open land on it. Asking you all to write. I must close for this time.

Affectionately, your daughter,

Myrtle

By haying time in August, Myrtle, Lewis, and Florence could pick twelve quarts of blueberries, gooseberries, and cherries in one afternoon. Lewis bought a cow at the Scandinavian settlement and kept her in the barn he had built a few months before. He and Florence put up about six one-ton stacks of hay, which work was broken up by chasing the cow, rolling fence wire, and canning cranberries and cherries.

Six men helped Lewis roll up the walls of a log house, eighteen by twenty-eight feet, in early September. He morticed the house, and got it ready for the roof. Five men came to raise the roof beams a few weeks later. By winter he and Florence moved out of the shanty into that solid place. He dug a well and found water at fourteen feet, which would save bringing buckets from the lake. When it rained, they could collect water that wouldn't have to be boiled before drinking.

One morning Lewis found his Jennie mare down on her side in the slough, and had to take a block and tackle to get her out. He doctored her, but she lay in the pasture until he had to use the block and tackle again to lift her to her feet. She was weak for several days, though eating, but died early one morning. The ground was too frozen to bury her, so Lewis had to build a funeral pyre around her.

Fishing was still good, and game that year included duck, rabbit, partridge, and pheasant. The potato yield was fair. When Lake Itasca froze, they went skating, and they heard a wolf howling at Fred Maw's one night.

A floor was laid in the house, and Lewis cut in two windows, built a milk cupboard, and chinked the logs. George Heinzelman, Lewis, and Mr. Johnson built a small cabin, ten by twelve feet, on Myrtle's claim. George and Myrtle went to town, a two-day trip, to get more lumber, roofing, stovepipe, nails, flour, and other supplies. Myrtle's place on Squaw Lake was ready by mid-November. Not finished, to be sure, but livable, with a good stove. Lewis's place was nearly done, but they had to tighten things up and prepare to leave for Owatonna. Lewis and Florence stayed with the Maws while Myrtle came back to Waukokee for Christmas.

Lewis and Florence joined us toward the end of January 1902. Warren's youngest, Kenneth, was walking and talking some by then and little Myrna was quite a musician for her age.

February brought the sad news that Rose's sister, Lucinda Ruanna Howe Mathews, died in Austin, at age thirty-five. A telegraph message came courtesy of a neighbor, emphasizing our need for a telephone. Rose was sick with asthma and rheumatism with Warren, his son Elvyn, and Myrtle feeling sickly too. Soon after, Warren's house was quarantined with scarlet fever.

Lewis was a big help cutting stove wood for me and for Warren. Florence sewed some new shirts for Lewis then started on a new dress for Rose. Even though it was fifteen below zero, Lewis drew up plans and started cutting wood for a bay window on the house. He and Sinland cut a hole in the house and framed it up, finishing almost half of it in one day. Everyone got vaccinations against the fever, but it made Florence sick and Lewis's arm got so sore he couldn't work on the window for a few days until the swelling went down. He and Florence spent several nights comparing the Sears and Roebuck catalogue with Montgomery Wards, deciding on their spring order. In late February, Lewis and Florence left for Owatonna again on the train.

He left the bay window for me to finish and it was not an easy job of shingling with its hips and kinks and corners, necessitating a great waste of time and shingles. After the sick were better we had the horrid job of fumigating. Sulphur burned in two rooms at a time, and more of the same medicine followed until the whole house from cellar to garret got it.

Myrtle started for Itasca on April 1st. On April 14th I went to the station to meet Cora Howe and Verna, our dear departed Lucy's daughter, who would be living with us.

Myrtle had taken a cat up to Lewis's claim, and he caught six mice the first night. We learned that George Heinzelman had been elected constable to help keep the peace with the lumbermen. Lewis wrote:

Dear Folks at Home,

Am at last seated to perform a much neglected duty — conditions previous to this have not been such that I could write since coming up here and while at Owatonna didn't know what to write, so waited. But today conditions are favorable, at least my stomach is full and a warm fire is in the airtight. Although it's not such cold weather, a fire feels cheery on a Sunday when there is not much to do. We have not said anything about Sabbath School yet for we haven't been settled well enough yet, but hope to go next Sunday, if weather permits.

We are expecting our goods from Sears on Wednesday, containing our stove, plow, gun, and numerous minor articles of utility. We make trips to the Settlement about every two days with our one-horse buggy and carry back all we can tuck in, which is a good deal. I usually walk back, and Lady draws my weight in truck. Florence will tell you what a trial it is to cook on an airtight heater, and I'll say she does bravely. Here's a synopsis of dinner today: boiled potatoes, fried fish, bread (baked on Myrtle's claim), milk gravy, tapioca pudding, coffee, honey, and sugar—how's that for high living? We waited a day between trips as the roads are too bad on the horse to go daily. She stays in fine order so far, and thinks the wide blade hay fine—eats it up clean. She drives on these rough roads splendidly, so careful among the roots and ruts that we are surprised at her. She will hardly rest in the middle of a hill but is eager to get to the top. She is a Hamiltonian and very much like the team at home for disposition. I get our goods drawn from Park Rapids, where they are shipped to for $6 a ton. The loads come to the store and we draw the small articles in the buggy leaving the bulky ones for a wagon.

We have been plastering the house inside with "northern grown plaster." I put almost a quart of ashes in each pail of mud and I never saw lime and sand plaster any harder when dry, and a light gray color—the dirt was not black to begin with, but rather was white clay and sand. I have the chamber and kitchen done and will do the sitting room Tuesday of this week. I made a temporary stairway of two-inch stuff which I shall use in the cellar after I make the other stair, which is waiting for new planks to come and will be a rainy day job.

Our farming season begins now in a few days as it freezes only one inch deep nights and we can work afternoons. Florence has begun her garden, and we have a fine lot of seeds from Northrup King & Co.; I sent a $1 order and Florence sent a $1 order and we each got a premium. The premiums are field corn, speltz, slender wheat grass, and soya beans. The beans resemble the butter bean only round. Speltz resembles barley, the grass looks like chess a little. I will give them a trial. Potatoes are 60¢ and 75¢ per bushel here. The little $13.50 cow I got at the auction gives us about a quart of real rich milk, night and morn, and the cow the store-keeper has is drying up. He thinks she will come in a month or so; if so, we will have milk continuously after this for a while, for the little cow is due to come in the 5th of September, the man I bought it of said.

We get mail three times a week now from the north. The mail man is to bring my stove and goods. I will be glad to get the plow and gun. A big flock of geese flew over our house the other day and I fired at them with the rifle, but was bothered by a neighbor's shell among mine. It was too large to enter barrel and I had to throw it out and throw in another, but by then geese were too far away.

Well, my letter has been about us and things up here, haven't asked a question, but all news will be eagerly read—about individuals, objects and operations. When are any of you people coming up here? How's Ma's asthma? Goodbye,

Ever your Son, Brother & Uncle,

Lewis

Work on their house consisted of laying the rest of the floor, peeling logs on the inside walls, putting up room partitions, and plastering with that mud mixed with ashes. Burning and clearing was the order on both claims, and planting asparagus, tomatoes, carrots, mangles, corn, squash, pumpkins, potatoes, rutabagas, beans, eggplant and cauliflower. Their cows gave more milk once they were on grass, and the hens were laying two or three eggs a day. Dolly had a heifer calf her own color, black. A setting hen came off with nine chicks.

— o —

Myrtle Lake

By 1902 Fred's journal picks up again, which is integrated here with a batch of letters between Myrtle and her parents. Letters tended to be long and detailed in the days before telephones, and give a glimpse into the daily routines and personalities of the writer.

News from up north reported normal happenings, while down here we were swept by storms and floods. The storms began violently on Saturday, May 17th, with continuous thunder and vivid lightning that struck and killed Harry Conkey while he was in his pasture attempting to catch a colt. He was thirty-four years old, survived by a wife and an eight-year-old daughter. The rain continued all night, creating flash floods that caused great loss of life and property. Three children of Mr. and Mrs. Herman Willbright of Forestville were drowned as the family tried too late to evacuate their valley, driving up a ravine when an immense body of water came rushing down a side gully. The water struck the buggy, overturning it and killing the horses. The angry torrent washed away their eight-year-old girl and two-year-old boy. The parents survived by clinging to the branches of a tree until the water subsided, but an infant girl died in her mother's arms before they could reach safety. Words were feeble instruments to express sympathy at a time like this.

Many in Preston were affected by the flood too, as the high waters filled the first floor of homes in the valley. Two buildings were moved back from the street one hundred feet and lodged against some large willow trees. The Weyhrauch family, who refused to get out after they had been warned, would have been carried away by the torrent if not for those trees. When the water went down it left a four-inch deposit of sticky mud on the floors, ruined gardens, and laid waste to most houses on the flat. The platform at the rail depot had been moved, the lumber yard was flooded, and considerable wheat stored in the mill was ruined. People were still cleaning up when another storm hit on Tuesday night causing more flooding Wednesday morning. The entire dam at the upper mill was destroyed, and six miles of railroad tracks between Preston and Harmony washed out, perhaps meaning the line would be abandoned.[*] In Waukokee, our grain and corn in the valley was gone, but the house was high enough to avoid the worst.

We sent these news reports to our folks up north and learned they had received lots of rain, too, but not as bad. Mosquitoes, though, got bad again in June up there and Lewis built screens for the windows, which also helped against the black flies. Myrtle sent us their news:

Itasca, June 9, 1902

Dear Mother,

I was so pleased yesterday at the receipt of your letter, but surprised indeed to hear of the havoc floods have worked down there. I don't get my paper at all, so all I know is what you told me.

I am very sorry you are having such a time with floods and storms. We did have some terrible storms here a week or so ago, but now our weather is quite fair. One eve I waded water at least fifteen inches in depth for about two rods right out on the high (usually dry) road.

When I began writing the children were all out, but now there are four small boys standing around my desk telling stories of rooster-fights, so my letter is liable to be a queer mixture. Well, I'll quit trying to write until this evening, the kids are chattering around so, and then I suppose I shall be too tired.

I am glad you are feeling so much better. I can't complain of my own health, but my work is very hard and I am kept so busy from eight-thirty until four-thirty that I haven't breathing time, and after that I'm almost too weary to write or do anything else. I've only been to gun club once this spring.

---

[*] *Preston National Republican,* May 23, 1902.

I was over to Lewis's last Saturday, but not down to my place as all were too busy to go with me and I haven't been brave enough to go alone yet. We've made a new road and I don't know the way very well. It is so brushy I shouldn't dare go without a gun, either. Two of my girls and boys saw a wolf cross the road ahead of them this morning, and when another party of the children came they reported being frightened by hearing wolves in the bushes growl.

The big horse flies are something terrible in my school house today. They bumble around about like the white-head hornets down there and bite right through a calico sleeve. Last Friday I smoked the mosquitos out of the school house in the morning and again at noon. Yesterday my little Vern ran away from school, so today I was obliged to administer a whipping, but not very hard. We have a dog in our neighborhood which persists in coming to school everyday. I'm going to put his name on the register pretty soon.

Will you celebrate the Fourth? Fred is going to have a picnic over by his place and we are invited. Do you ever hear from Grandpa? Where is he? I can never remember whether it is Thetford Hill or just Thetford. I am going to write to him someday if I ever can think straight long enough at a time.

Tell Warren that I think he could have this school if he should apply for it, as I am through with teaching after this term. Of course, they can't hire until after school-meeting. I guess I've written everything I can think of so will not try to keep you reading trash any longer. If you have any more flooding notes from papers just send them along as they are very interesting, to say the least of them. Affectionately,

Myrtle

During harvest time both Florence and Myrtle got sick, so George and Lewis had to do triple work haying at Lewis's, Myrtle's, and George's claims. Cranberries were plentiful, sweet corn was harvested, but the cow ate up corn saved for seed. Lewis got another job at the settlement working on the school house putting up wainscoting, floors, and a window. A camping party came by and paid $1.50 for breakfast.

In September, Myrtle was a delegate to the Minnesota Congregational Association Convention in Fergus Fall, and stayed with a nice young couple, Dr. and Mrs. Drought. She traveled to St. Paul and Wadena, too, and had some eye and dental work done along the way.

On election day, November 4th, Lewis served as clerk of election working from nine in the morning to ten at night for $3.80. Rabbits were abundant that year, and Lewis got his first deer, a doe. Florence and Myrtle took in washing for the lumber camp, earning 10¢ per piece. Florence made $1.10 and Myrtle $2.50 the first day, and the next day they got the washing for eight men.

Itasca, Dec. 7, 1902

My dear Mother,

I've been wanting to write you since Thanksgiving but have been very busy. Thanksgiving day in the afternoon I went over to my house and took Fudge, the big cat, with me for company and intended to stay that night. I had just got rested a little, brought up some water, shot and dressed a squirrel, when Mr. Korth and Mrs. Heinzelman came over after me so I locked up again to go with them.

Mrs. H. had a crew of men who were working on a new dam—a sluice-way dam just below the old one—boarding here and she had a "Cookie" from the camp helping her in the house, so wanted me to come over just for company. I came, but that eve when Mr. Cookie went after the mail he also brought home $2 worth of whiskey so the next day he wasn't worth much and left in the afternoon. Then I put on my big apron and went to work and we did the work and cooked for fourteen men—sometimes eighteen at dinner, for Mr. H. had surveyors here working on the town site.

We got along nicely with the work and didn't get very tired until the last day. They went away Friday morning and we did a big washing on that day—thirteen washerfuls—and yesterday took life kind of easy. Today is very cold—twenty below zero this morning early—so I did not go to church as I have a cold. Lewis and Flo didn't come over to church either and I suppose the reason was that they didn't dare to let the house cool off for fear of freezing the potatoes in the cellar.

Heinzelman & Korth's Logging Camp keeps the men hustling night and day. They have about twenty-two men in camp and want a few more, which Mr. H. expects to get in Bemidji, and then they will get through logging in about a month, I guess.

Alf Ayers left Lewis's Friday for Mallard Lake, thence to Crookston where he expected to meet Fanny and they each would file on a claim. The claims are about seven miles from Lewis's and it will be hard for them to live there, but it was the nearest they could get quarter sections adjoining. And, between you and me and the rest of the family, it is near enough for comfort and pleasure.

We had lots of excitement while the "dam men" were here, as Mr. Wood was among them and this time he wasn't intoxicated. He didn't act as silly as before, but he won't speak to George when he meets him. Bill Diamond was foreman. They were here three weeks and a very agreeable set of working men, hand-picked from several different camps, and the work and boarding place seemed to be appreciated by them all. They were always gentlemanly and sociable, but never in the house excepting meal times. They had bunks in the summer kitchen.

Lumberjacks are generally such a disagreeable, bummy set that we feel like making as many bright spots in the lives of the decent ones as we can. The lumberjack question is an important one among pastors in this part of the world.

They usually spend their Sundays playing poker and gambling and the preachers visit the camps as often as they can and leave magazines, etc., for them. One Reverend Higgins, a Presbyterian, has given up pulpit work for camp work. He works among them all the time, in every way he can. One man, a drunkard who had a family, was raving when without drink. Rev. H. bought whiskey for him and got him, while pacified by it, to go and take the Keeley Cure. For that, he was censured by a great many, but the man was cured. We can't even imagine the other compromises that minister makes to get on the "good side" of the lumberjacks.

I dread the time when the name Itasca shall suggest a town of saloons, as a great many predict. It seems an utter impossibility for a town in lumbering regions to be any different, but it is awfully hard for me to think of. If it could only be as nice as its name and have attractions worthy of the beauty of the place, how the people might enjoy it! But I expect to be heartsick over it a great many times.

If the lots sell well, there will be a great demand for carpenters in the spring and I believe Warren could work in well. I shall watch the prospects as closely as possible. George is going to have a business lot and a residence lot. The first lots will sell quite cheap—from $25 up.

I will help Mrs. H. get dinner now. We are going to have chicken for dinner and Mr. H. will take dinner at home today. He gets very tired of camp life.

This is a beautiful day. We have just snow enough for sleighing in the woods now but not in the sandy roads very good, although Mr. and Mrs. H. are going to Bemidji with sleigh tomorrow and I will stay with Miss Rose until they return. Now I must go to work so goodbye. As ever, your daughter,

Myrtle

I went to town and got Myrtle's glasses, which she sent home for safety. The bows were gold and cost $7 at one time, but she outgrew the lenses and got a new outfit costing $15. We also received some beautiful pictures of that north country, which we highly prized and expected to frame. Our own pictures were also finished and at home, and they were quite satisfactory. We sent Myrtle a picture of us and brought home a new clock from Conkey's to be paid for in wood. Later, I went back and exchanged clocks at Conkey Bros.—the first one didn't work right. I took about a half cord of wood to the firm also as part pay for same, and returned the next day with more wood, making nearly one and a quarter cords to Conkey.

Two men came to my home and accused my dog of being one of a mob to kill fifteen of Bremseth's sheep I killed the dog, but then I was expected to pay the loss though the owners of the rest of the gang could not be found, so there was a question of adjusting the penalty. No one knew how many dogs there may have been in the scrape and I didn't know as the law would make that a point. Two weeks later, Alex Long came here and settled the sheep-killing scrape by demanding ten sheep, which I gave him, still protesting that my dog never killed a sheep.

I butchered a hog weighing 155 dressed and made 40 pounds of it into sausage. A beggar came, the first in a long time. He looked woebegone and needy enough. I gave him 20¢ to help pay his passage back to Germany, and advised him not to get drunk on it.

A blizzard started on January 29, 1903, which left the roads icy, bare, and well nigh impassable, especially for an unshod team. Our water supply grew less by degrees and I worried that it would stop before spring, owing to the last year's floods digging out below the spring, draining the water off. I took off the upright pipe to keep it from freezing up entirely. For a week we had kept the water running by wrapping it in the nighttime and pouring hot water on it occasionally when it stopped running. By the end of February our water famine was the real thing, and we had to depend on melting the snow.

I cut down the old oak and maple that stood south of the bay window because they shut out the sun and light. They were at least three rods away, but it was an improvement. There were still two butternuts and a beautiful elm there yet. The old oak mowed the top off a maple when it fell, which I hated to lose, and I coaxed it to live.

The bluebird and robin came March 8th. There was a town meeting, and I went on horseback in fog and mud. We voted on a proposition to change the road system from the pathmaster method to another, thereby abolishing the poll tax arrangement. It was a sort of town option that had been in vogue for several years, but our town had never ventured its trial before.

I set a row of little maple trees from the garden along the road to the crabapple grove. My apple trees were almost ruined by rabbits, so I had to graft them. I took a hog to town that weighed 180 to my surprise and joy. I got $6.35 per hundred, which relieved my stress for a time. The wind was strong and chill coming back, and I took cold as I had left the warm coat home.

George Kaasa boned me to buy his colts. I traded for the colts as follows: I gave seven sheep with lambs and the old buck, and my note for $50 for one year at six per cent. My flax went to market, five and a half bushels, and I got $4.30 for it, less than it was worth the previous fall by 8¢ per bushel. I took advantage of a cold but pleasant morning to tag my sheep. A job quite necessary but decidedly disagreeable.

***

We heard news of a horrible murder in Granger. Neighbors had noticed that one Mr. Krueger had not been seen in town for a few days, and finally notified the sheriff. Mr. Krueger's "house-keeper" stated she did not know where he had gone. But it was soon discovered that Krueger's woman had sought refuge two miles out of town during a big, howling storm late at night the week before. After searching around the valley for several hours, Mr. Krueger's body was found partially submerged in the river. A rope with a weight had been tied around his body and when they pulled him out of the water there were grevious wounds on the body.

The late Mr. Krueger was a decent, upstanding citizen who, after losing his wife a while back, decided he didn't like living alone and needed a female companion. He saw an ad in the paper: "Decent, respectable lady, age thirty-five, would like a position as a housekeeper for a middle-aged bachelor." Krueger answered the ad, and she came; a sturdy-looking woman from Wisconsin. The following investigation revealed that the housekeeper had killed two others before. She had hacked Krueger to death with a pick ax, then pushed him into the flooding Iowa River. Had it not been for the loud clap of thunder that startled her horse, breaking the wagon hitch, she might have made it back to the house with no one the wiser. Prison was her last home.[*]

***

Old Billy seemed unable to work and was taking a rest, but my corn ground was only half prepared and I hoped against hope. I planted early potatoes and some squashes. A beautiful Sabbath came mid-month and I felt the need of it. I had built fence for four

---

[*]   Orell Selland, *Memoirs and History of Granger* (Harmony, Minnesota: Big Woods Graphics, 1987).

days and was worn and weary. Driving posts and dragging wire was heavy work; perhaps a more practical man would have used a horse, but I was afraid of tangling up and spoiling the wire. I built 105 rods of fence, 95 straight rods and 10 across the north end.

A heavy rain brought higher water, and still another at midnight flooded the fields. These rains were not common summer showers, but real cloudbursts downpouring floods from the sky and trying my nerves to a point almost unbearable. Nothing was safe, and ruin was on every hand. I bore up last year, and tried to take my medicine this time, but I realized there was an end to human endurance and we were very near it.

I found a dead sheep in the pasture, cause unknown, but I suspected dog work. Warren's school was done, and we got a letter from Myrtle, along with some fish.

Itasca, Minn.
May 29, 1903

Dear Father and Mother, Brother and Sister,

Some few days ago I sent a box of smoked fish to you people and think perhaps you have received them ere this. I am "at home" and have been all the week getting some clearing done. I expect I shall be needing my filing papers pretty soon as I am thinking pretty seriously of proving up. Since I will have met the one-year residency requirement, I can buy out my claim. This will reduce my savings, but will be profitable in the long run, I hope. I have to advertise my intentions five or six weeks previous to proving up and probably won't get a deed for two months yet, so keep this among the older members of the families of Cummings, lest it should go awry. I will write a separate letter for the children to read with most of the news in it.

George has two claims and has decided (with my help) that he will prove up on the one he won't have to pay taxes on after I prove up on mine, and then I will help him hold down his remaining claim until his seven years are up. I am sorry to leave mine and move to his on Lewis's account, for I know they would like us as neighbors, but George's is so much better farming land and so much more handy to the settlement and store. Plus, he could not afford to by a team or to buy them feed which would have to be hauled way over in here. So we decided to locate there, where he can exchange his work for team work when needed, and we won't have to haul everything we have to eat nearly so far, as his house is not more than half a mile from the store.

If he had not lost on his logging deal last winter we could come home next fall. He says I shall go anyway, but I will have to think about it. I know you are anxious to see and be acquainted with him. Anything I can say of him is seen through my eyes and of course is good, but I never have known anyone else to say anything bad of him and I can't imagine what it would be. He is strictly honest, temperate, hardworking, good-natured, kind-hearted, and a Christian. He has been pretty patient "baching" it up over there so long and it is very inconvenient for us both.

He is clearing now while I am sitting under a tag-alder bush on the edge of the piece. Now I guess I'll wind this up and say the rest in the other letter. I hope the fish will be good when they reach you and perhaps we can send some more some time. It isn't much of a job to smoke them. When you write again I guess you had better send my filing papers, if you please, and then I'll have them when I want them. Since I've decided to prove up on my claim I can't teach, because I must stay at home for the duration. George is helping hold camp. Good bye,

Myrtle C.

I finished potato planting and went to town for seed beans and to mail Myrtle her filing papers. A letter from Vermont announced that Father had reached his 90th birthday, and that he lived with cousin Tim Cummings in Aunt Lydia's old home, where I lived between my mother's death of tuberculosis and the time Uncle Fowler brought us here. A card from Rose's sister, Mary Burns, told us that she became a widow on June 20th, as George Burns died that day from kidney disease.

July 1st tried to clear away after a storm of two days with rain enough and plenty of the other entertainment usual to such times. I rather suspect that other localities got a worse handling by the way the clouds were tossed about. The full moon brought a big downpour in our locality, with terrific electrical disturbances and worse in some other places. Cloudbursts and landslides down by Hokah stopped trains for several days.

A bold, bad wolf celebrated the Fourth on my sheep, killing two, wounding three, and one of the latter died later. My haymakers began work the 13th and were half done when we had a soaking rain, which of course did much damage to the hay. The corn struggled upward and rust was on the oats, but potatoes were quite thrifty.

July 19th reminded me that this date, thirty-five years ago, I first met my wife, "A beautiful Rose in the back woods." I set some posts up the ravine to the spring to change the sheep fence to let them down near the house and make it handier altogether and all round, and I don't know why I didn't plan it so at first. I went to town with Warren early; institute for him, to have an ulcerated tooth extracted for myself, and to mail letters to Mrs. Gould and Myrtle.

My oats were all nicely harvested and shocked by August 7th, although crinkled terribly, every straw was cut and saved. We were treated to a driving storm the first of the week and I felt sad to see the result, and was afraid it never would be harvested, so when Ev Miner came with a patent device for lifting the lodged grain I was surprised.

Eight small pigs came by surprise at night when Rhiel's stock broke into my yard. Mosquitos and flies were very big. The wife went to a church sociable and came home sick with a colic that they had to sell there—most of the guests had it more or less, and some were very sick.

A letter from Myrtle reported an elopement at Owatonna too disgraceful to mention. Not the first we had heard of it, but we were forced to accept the rumor and to acknowledge the black sheep, Florence's brother, as "relation." Fred Maw left his wife, Carrie, and their four children in Owatonna, where they had been visiting her mother. We learned later that one of Myrtle's neighbor women, who was supposed to be visiting her mother, actually disappeared from Itasca about the same time. Mr. Frankenberry, the abandoned husband, tried to track them down for more than six months without success.

The frame of my spring house was complete, ready for lumber to cover and side it, which put me to my trumps to get. I went to Preston for lumber for the roof and bought 1,250 cedar shingles and 140 feet of rough lumber. Dr. Phillips was here to see Jennie and gave me some "nerve medicine." I worked on my roof and nailed on 750 shingles.

A little boy was born to Warren on Aunt Lydia's birthday, August 26th. The sun shone out between the flying clouds after a week of dark and wet. We did very little but care for the sick and tend the new baby. I had a roof on the new building, but the north side of the wood shed needed a new roof too, which was my next job. We received a letter from Lavina announcing a daughter born to her daughter, Lillie Waumit, on the evening of August 27th, which meant Warren's little boy has a second cousin, I think.

I finished shingling the wood shed with cedar and got some ship lap to side up two sides of the spring house. A week of too wet to stack, and my oats were settled down so much that I feared they would rot. I got my spring house enclosed and the wood shed shingled. Another wet week, but I managed to haul out some manure, and brought up some gravel from the creek to grade around the new milk house; seven loads in all, and I wish I had as many more.

A letter from Myrtle came, full of hope and plans of future in which matrimony played a very conspicuous part. Kennie, Warren's three-year old, made it interesting for us one eve by losing himself in the woods, but his grandma found him after a long search.

October 3, 1903, was a terrible day for the city of St. Charles, just twenty-eight miles north of Preston. All that is necessary to keep it in memory is the date and the one dread word: cyclone. The death list was eight, with scores of wounded, and the town was practically wiped out of existence. Preston escaped with only torrential rains and a stiff gale, though the clouds looked threatening. The northwest corner of the county, near Spring Valley, was hit pretty hard, damaging area farms and destroying the old landmark stone church at Sumner Center.*

I went to Preston with four posts for Mrs. Weybright; flax, 130 pounds for market; and the first installment of butter for Professor Lurton. We furnished him five pounds every two weeks thereafter. Our first butter sold since we quit the creamery totalled thirty-three pounds at 18¢ per. Old Major Hotchkiss, being eighty years

---

*   *Preston Times,* October 7, 1903.

of age, laid aside his pen and sold his outfit after fifty years of labor at the editor's desk.

Myrtle finally owned her claim up north, and the park ranger named "Myrtle Lake" in her honor. The lakes and ponds were numerous up there, surrounded by timber land, and they were good for fishing and bathing. Earlier in the year I dubbed our little farm "Fairy Glen," which I felt captured the neatness, charm and quaint feeling of our valley. But, as the poet said, "What's in a name?"

I finished digging potatoes by October 22nd and had about twenty bushels. Potatoes were three-fourths ruined that year: rot, rot, rot. I think three bad ones to one good was a conservative count. Wolves were bold and noisy, and rats and mice innumerable. A flock of geese flying overhead brought us the news of cold weather up north, and the wind sang its weird autumn anthem around the corner of the house reminding us of unbanked cellars and tumble-down stables. I worried I would have to sell some of the cows for lack of feed, but decided to keep them awhile yet.

We received a letter from Myrtle in which she fixed her wedding day as November 15th, and requested us to send her the material for a dress to be used on that occasion. We went to town the next day and got the desired goods and mailed them to her.

Itasca, Oct. 21, 1903

My dear Father and Mother,

I am going to write "in a hurry," as I always say, and probably won't say very much, but will come to the business part of it first and later will do the best I can at a letter. I have decided that I can't do better than to order material like the sample you sent and as it is wearing along toward the time, I begin to feel a little anxious to be at work at the dress. So if you will go to Hard's and get nine yards of that Henrietta and four one hundred-yard spools of silk to match the goods and send it as soon as you can I shall be greatly obliged and will send you the money for the purchase. I would send it in this letter but I will not have it for a week or so until I have payment for the camp job.

We've just been picking out our little furniture outfit and it amounts to $39.21. It consists of an iron bed, hotel stand (dresser and commode combined), oak extension table, reed sewing rocker, bed-springs, dinner set, washer, wringer, Acme cottage heater, and a mattress. George has one nice iron bed, a mattress and some bedding. I wish you folks had a Sears, Roebuck and Co. catalogue so I could tell you the numbers of the things we ordered. Good-bye for now,

Myrtle

October 27, 1903

My dear children,

I have tried and tried to find a chance to write ever since your letter came, Lewis and Florence. Now I can only write an excuse, for it is after ten o'clock and I have just finished my evening's work by packing a jar of butter. I work butter four times a week, we churn twice a week, and each mess of butter has to be worked twice, besides my washing, ironing, baking, housekeeping, caring for milk, and other things. The days are getting so short it keeps me busy every minute. My fall house cleaning has just begun and I don't know when it will end. I got things moved around and set up the stove in the setting room yesterday. I have quite a lot of sewing to do and when I get up in the morning I don't know whether to sew or clean house. My canning is all done but a few ground cherries. I have just 100 quarts sealed up, besides a three-gallon jar of sweet apple pickles. I am feeling quite well except my hands, as butter work is hard on them. I am glad to hear that you are feeling so well, Florence, and sorry for you both that you will lose Myrtle for a near neighbor, it will make it lonely for you.

Warren's folks are well and the children are going to school. Kenneth says "I am Grandma's boy now 'cause she found me up in the woods." The baby is good and grows nicely—he is quite talkative for a little fellow. He doesn't look much like the others, he has such black hair and heavy eyebrows.

Now I must just say a word or two to Myrtle and then quit, for the clock has just struck eleven. Your Pa has said all there is to say about your affair. I am pleased to hear of George's potato crop being so good. What are they worth up there? If it wasn't so

. far away we might try to get some of them for seed. Everybody's potatoes rotted so bad around here. I am pleased with your idea of a wedding day, it was Sunday when we were married.

—Ma

P.S. When your dress is cut out, send or bring us a piece, I forgot to show it to Pa, as he was in a hurry to get home. I think your brown dress is nice, Myrtle, as well as durable. I am making a red one for Verna, and will send a small piece. She looks best in red, and has a red worsted one that I will have to let down for this winter.

November opened fair and mild with a sort of smoky mist on the air, especially at night. The heavens were lit up with an "aurora" of some sort which was not at all to my liking, for I supposed it indicated some unusual disturbance in nature, boding no perceptible good to earth.

The sad story from Preston came that a saloon keeper fell onto a sharp stake from a grain stack and was killed. Frank Bakey was helping his father-in-law with threshing. One setting had just been completed, and the machine was about to be moved to another setting. Bakey had climbed to the top of a small stack with two or three others to better observe the construction of the train track in the valley below. All slid off the stack, the others without injury, but poor Frank slid full onto a stake about six feet long which was resting up against the stack. The stick entered his leg about three inches below the groin, tearing the fleshy part, and penetrated his stomach a full fourteen inches. He was literally impaled and, as he fell forward, the cruel stick forced its way towards his backbone. The unfortunate man was carried to a nearby house and a doctor sent for, who remained at the bedside and alleviated the suffering by administering anesthetics until the end came at three o'clock Sunday afternoon. His shocking death cast a shadow on the entire community. His kindly, generous nature had made him a host of friends. He was ardently fond of music, and the cornet band would especially miss his talent. Born in 1868, he was survived by a wife and a four-year-old son.[*]

---

[*] *Preston Times,* November 4, 1903.

The railroad was completed from Isinours to Preston a few weeks later. The ground was frozen and whitened with snow, but I managed to plow another half acre, rolling up frozen lumps continually, making it very hard work for man and beast.

November 15th — thirty-five years before was our wedding day, and in 1903 the day was set for Myrtle's wedding. At such distance, however, we could not know whether it came off on scheduled time until a letter came from Myrtle dated the 17th, this being the announcement of her marriage to Mr. George Heinzelman after an acquaintance of two and a half years. She had always thought him a "nice young man" and had been in real love with him for two years, if appearances counted for anything.

Fairy Glen, November 22, 1903
To Myrtle C. Heinzelman
From F. A. Cummings

Dear Child,

I suppose I shall have to quit addressing you thus. We received your letter last week announcing your marriage, and were happy to know that you were so quietly and sensibly started on your life partnership with the one of your choice. You say you are sure he will be kind and good to you, and I am just as sure that you will be kind and good to him. Now I wish you both a lifelong season of joy, such as only virtue and unselfish love is worthy of. I feel much satisfaction in looking over your history for you have accomplished something that few girls have had the tenacity to obtain in holding that claim in the wilderness until it is yours indeed. Now I hope you will still appreciate its value and hold on to it and enjoy the fruits of your labor. You haven't told us yet about your new home, but I presume you will in the near future. How far will you live from Lake Myrtle and in what direction? I hope that it may always be called by that name while it has a pailful of water in it to commemorate your victory. Of course you have had lots of help from friends who have stayed loyally by you, and to them is much of the praise due also. But they no doubt saw something worth their trouble in my "little girl" and of course that makes me thankful and happy and — well — proud.

November 29th, Warren's people are living here now and have measles just to spice the dull routine of life. Kennie has had them and is all well and we are waiting now for the others. Don't know whether Verna has had them or not, but hope so. We are all right otherwise. I have been working night and day to get my corn out and this accounts for the delay in writing. I had to stop plowing due to frost, with two more acres to plow. I now have about three and a half acres of corn to husk. Our water works are playing out again this winter, and I am too busy to look after it as yet.

Now, as to the butter, we will fill a two-gallon jar as soon as possible, box it up and send it by freight to Park Rapids. It will probably take two weeks, and we will send you a letter when shipped. Preston has a new railroad now, so that we can ship from there. We have a contract to furnish five pounds to Preston people each week so that it will take longer to fill yours. Your Ma will add her note, and we will go to the post office tomorrow.

Waukokee, November 29, 1903

My dear Myrtle,

I will try after a long delay to answer your letter received over a week ago, but the measles kind of broke us up the past week. Warren can't come home every night very well, so I told Jennie to stay over here so we could help her take care of the children. She and Baby and Kenneth have occupied the sitting room, with the crib, cradle and cot-bed, and got along quite comfortably. Warren's health is real good and things don't seem to worry him as much as they used to, he is more like himself than he has been for several years, and he thinks so much of his family. They have everything to make them comfortable. Warren and Jennie have just decided today to name the baby Vinton Augustus, he is a sweet little thing, good as can be. Pa says he looks like Myrtle used to, he plays with his hands and sucks his finger, laughs and coos as babies usually do, and he tries to sing some.

I presume you are settled in your new home with all your nice new things; I should like to see them. I wish you all the joy that it is possible for you to claim in this world, as well as in the next, and I congratulate your husband, for what he has gained is my loss and I feel it deeply. It begins to look as though you will not come home this winter, but I hope it will not be very long before you can. I want you to have one of my feather beds and I want you to take your choice in them. I have some nice little dishes that I have been saving up for you, but you have a full set now, so I suppose you will not need them. I think you got a nice outfit to begin with cheap enough. I feel very much better to know that you are settled and have a protector, than I did when you lived alone in your cabin, although you seemed to enjoy it and you will have it to look back to that you secured it for yourself. But what hurts me the most is that you are so far away. I had always hoped that my children would live near me while I lived. Our birthdays and anniversaries are the same, but I am afraid we will not celebrate them together very often. Jake Davis and family visited with us on our anniversary day, and Armada remarked that she had never seen anyone grow so gray as fast as I had in the last two years. I told her I thought it must be because I was lonesome.

We reorganized our "Lady's Aid" four weeks ago. We met at Armada's and we have thirteen members. They nominated me for their president, but I declined to run, so we elected Miss Margaret Durst; Orissa Engle is vice president, Armada, secretary, and Katie Durst, treasurer. Our plan is to serve supper where we meet every two weeks and each pay 10¢. We may think it best to do some work after a while, but have not decided yet.

It has been snowing by spells all day and we now have about an inch. The first we have had to cover the ground, as we have had fine weather nearly all fall.

You said people up there were surprised when you were married. It was not so here, everybody seemed to know it a long time ago. Every place I went this summer someone would say "I heard Myrtle is going to get married." Well it is nine o'clock and everybody is stirring around getting ready for bed so guess will have to close soon. We will look for a letter from you again soon and you will hear from us when the butter is ready to go.

Good-bye, with love and best wishes. Give my regards to your husband.

From your loving Mother

Our water supply ceased December 1st, and then we brought it in pails from the spring. The measles dragged their slow length along and our home was turned into a hospital, making lots of extra work and a disagreeable time for us generally.

I sent some butter to Myrtle by freight from Preston. The weather was cold, blustery and growing worse every hour, ten below at sunset. We had about six inches of snow, light and loose on dry ground. The measles hospital was finally broken up and the patients moved back home. We received a letter from Myrtle with a picture of her husband and a contented expression of hope and satisfied ambition. A sort of "all's well," so to speak, with just a little homesick feeling.

I finished husking on December 23rd, a most beautiful day. Our church people were soliciting aid to buy Brother Jones a Christmas overcoat. We gave $1 toward the enterprise. Christmas Day was twenty-two below zero, with much wind.

We went to town with eighteen pounds of butter and eighteen pounds of chickens. I bought a duck coat and overalls. A steer was butchered here, and I sold Albert Iback two pigs to be weighed and settled for later. The beef weighed 545 pounds, and Warren got half of it. This finished up the year's record, and the last day was fine, ending with a full moon within a circle. The old year left a horror hanging over Chicago—600 lives lost in a burning theater; men, women and children perished in a pile.

—o—

Epilogue

There are two Myrtle Lakes in Itasca State Park, and research suggests that our Myrtle's lake is in the northwest corner of the park. Securing the homestead claim was a significant milestone for Myrtle, as was Fred's expression of pride in her achievement. He could not have been happy, however, at her suggestion that Warren might like to come up north to take over her school. Warren stayed near Waukokee though and, eventually, Myrtle and George moved back there too.

Fred lives on to the ripe old age of ninety-two and continues to keep a journal almost to the end. He records news events of the day ranging from World War I to the Lindenbergh kidnapping. He watches his grandchildren become intrigued with automobiles and new-fangled farm machinery. Outliving Warren and his beloved Rose, Fred leaves his writings to Myrtle, my great-grandmother. I remember meeting her as a young child and hearing family stories about how she rowed across the lake to get her mail. But I never really knew her, or Fred, until now, and I am so, well – proud.

— o —

Glossary

ague — a fever, usually malarial, marked by regularly recurring chills.

apoplexy — stroke; sudden paralysis with loss of consciousness and sensation.

appurtenances — apparatus, equipment, accessories.

avoirdupois — great weight.

a "bee" — a social gathering to assist; a meeting of people to work at something together.

Belial — wickedness or worthlessness as an evil force; Satan; one of the fallen rebel angels.

biddy — a hen.

blackguard — to abuse with words; rail at; revile.

broke land/breaking land — to tame or make usable.

catarrhal affliction — inflammation of membrane in nose or throat, causing an increased flow of mucus.

cavil — quibble; to object with little reason.

chess — a grassy weed.

cholera infantum — intestinal disease of children.

clevis — U-shaped piece of iron with holes and a pin to attach one thing to another.

diphtheria — an infectious disease caused by bacteria, causing weakness, fever and obstructed breathing.

epizootic — an epidemic among animals.

foment — stir up, incite.

freshet — a sudden overflowing of a stream due to melting snow or heavy rain.

froe — a wedge-shaped cleaving tool.

grippe — influenza

grubbed — to dig up by the roots.

harrowing — dragging a heavy frame with spikes or sharp disks to break up and level plowed ground.

havoc — great destruction and devastation.

Hibernian — an Irishman.

hob — to make trouble for; interfere with.

hydrophobite — fears of water; unable to swallow water; not capable of uniting with water.

hypoxia — a condition resulting from a decrease in the oxygen supplied or utilized by body tissue.

jingo — a person who boasts of his patriotism and favors an aggressive, warlike foreign policy

John Brown — Abolitionist who led a raid on an arsenal at Harper's Ferry and was hanged for treason.

Josephus — Jewish historian c. 37–95 A.D.

kine — cows, cattle.

La Grippe — influenza.

lodged — to beat down growing crops (as, by rain.)

lowery — dark and cloudy.

lyceum — a hall where public lectures or discussions are held.

mackinaw — a heavy wool cloth, usually with a plaid design.

mangles — beets

Medes — natives or inhabitants of the ancient country now known as Iran.

milk fever — a disease of dairy cows, often occurring after calving, sometimes causing paralysis.

milk leg — a painful swelling of the leg caused by inflammation and clotting in the veins.

Modocs — Indian tribe formerly occupying territory in southwestern Oregon and northern California.

nihilists — believers in the doctrine that all existing social, political and economic institutions must be completely destroyed in order to make way for new institutions.

non compos mentis — not of sound mind; mentally incapable of handling one's own affairs.

peck — eight quarts

phrenology — a system of analyzing character by studying the shape and bumps of the skull.

quinine — a bitter, crystalline alkaloid extracted from cinchona bark, used for treating malaria.

refractory steer — hard to manage, stubborn, obstinate.

riffle — a stretch of choppy water.

rod — a unit of measure equal to 5½ yards.

roily — turbid, muddy.

scarlet fever — an acute, contagious disease characterized by sore throat, fever and scarlet rash.

School land — land sold by the government to settlers, in order to fund establishment of schools.

scrape — a disagreeable incident; embarrassing situation; a fight or conflict.

scraped acquaintance — managed to get by.

Sharon — Coastal plain in West Israel extending from Tel Aviv to Mount Carmel

shock — stack bound grain on end, to dry.

sine die — without day, indefinitely, no set end date.

slue — to turn or swing around.

smallpox — a highly contagious viral disease characterized by fever, vomiting and pustular eruptions.

spotted fever — any of various fever diseases accompanied by skin eruptions.

Tam O'Shanter — main character of Robert Burns' poem by the same name.

tomfoolery — foolish behavior; silliness; nonsense.

toper — drunkard.

trow — believe.

typhoid — an infectious bacterial disease spread by ingesting food or water contaminated with excreta, causing fever and intestinal disorder.

verdure — the fresh green of growing things; vigorous; flourishing.

vinca — periwinkle.

viz. — *videlicet*, that is, namely.

wether — castrated male sheep.

yellow fever — infectious tropical disease transmitted by the bite of a mosquito causing fever, jaundice and vomiting.

INDEX

A

Adams, J. Q. 50
Arthur 147
Ayers, Alf 198, 239

B

Bagnell, Isaac 10
Baker 163
Bakey, Frank 249
Barnes, Rose 202
Beecher, H. W. 89, 116
Beltrami County 228
Bemidji 226, 239–240
Bowden 29, 33, 119, 121, 125, 128–
    129, 153, 158–159, 163,
    169–170, 192, 198–199
Brady 162
Bremseth 240
Bristol Grove 45
Brown, John 15
Bruce 202–203
Burns 140, 244
Burr Oak, Iowa 6

C

Camp Creek 14
Canby 57
Captain Jack 57, 59
Carnegie 127–128, 169, 204
Chicago 44, 62, 70, 94, 116, 168,
    199–200, 203, 207, 221, 225,
    253
Cleveland, Grover 157, 168, 188,
    192
Colburn 69–70, 112, 121–122, 169,
    191, 193
    Abner 167

Florence 69–70, 112, 121, 139
Nathan 69
Conkey 91, 140, 153, 156, 163, 176,
    185, 235, 240
Cottonwood County 31, 33, 51–52,
    61–62, 102, 137
Crosby, Charley 10
Cummings
    Asa 51
    Caroline (Caddie) 85
    David 29–31, 71–78, 126–127,
        129, 174, 195–196
    Elvyn 202, 231
    Eunice 71, 73–78
    Frank (B. F.) 21–22, 29–30, 44,
        83, 85, 91, 94, 99, 107,
        121, 127, 137, 153, 156,
        166, 186
    Harriet 51
    Henry 183, 197
    Jennie. *See* Jennie Taylor
    John 22, 75, 102, 107
    Kate 31, 83, 85, 94, 139–140,
        183, 188, 193
    Kenneth 231, 246, 248, 250–251
    Lewis 53, 64, 82, 93, 103, 161,
        163–164, 168, 171, 174,
        178–179, 181–188, 191–
        192, 195–198, 200–201,
        203, 215, 224–232, 234,
        236–239, 243, 248
    Marshall 138
    Moses 13, 15–16, 22, 31, 33, 44,
        52, 74, 82, 107, 138,
        156, 200
    Myrna 193, 195, 197–198, 228,
        231
    Myrtle 3, 101–103, 119, 122,
        131, 163, 172–173, 175,
        181–185, 188, 191–192,
        195–197, 199, 201–204,
        215, 224–225, 227–233,